The Business of Biodiversity

WITPRESS

WIT Press publishes leading books in Science and Technology.
Visit our website for the current list of titles.
www.witpress.com

WITeLibrary

Home of the Transactions of the Wessex Institute, the WIT electronic-library provides the international scientific community with immediate and permanent access to individual papers presented at WIT conferences. Visit the WIT eLibrary at
http://library.witpress.com

The Business of Biodiversity

Mark Everard

University of the West of England, UK

WITPRESS Southampton, Boston

Mark Everard
University of the West of England, UK

Published by

WIT Press
Ashurst Lodge, Ashurst, Southampton, SO40 7AA, UK
Tel: 44 (0) 238 029 3223; Fax: 44 (0) 238 029 2853
E-Mail: witpress@witpress.com
http://www.witpress.com

For USA, Canada and Mexico

WIT Press
25 Bridge Street, Billerica, MA 01821, USA
Tel: 978 667 5841; Fax: 978 667 7582
E-Mail: infousa@witpress.com
http://www.witpress.com

British Library Cataloguing-in-Publication Data

A Catalogue record for this book is available
from the British Library

ISBN: 978-1-84564-208-2

Library of Congress Catalog Card Number: 2008933932

*The texts of the papers in this volume were set
individually by the authors or under their supervision.*

The paper used in this book is FSC accredited, chlorine
free (ECF) and sourced from well managed and sustainable
forests. Manufacturing accreditation: ISO 9001, ISO 14001

© **Mixed Sources**
Product group from well-managed forests, controlled sources and recycled wood or fiber
www.fsc.org Cert no. TT-COC-002303
© 1996 Forest Stewardship Council
FSC

Printed by the MPG Books Group in the UK

Contents

Foreword by Jonathon Porritt

These days, the 'business and sustainability' agenda is fast-moving. Business finds itself absolutely on the front line of the battle being waged between humankind (as the dominant species on the planet) and the rest of the living systems and creatures with which we share the planet. This will be seen in retrospect as a collective aberration of monstrous proportions (we are, in effect, making war on ourselves), but it is taking us a very long time indeed to wake up to the consequences of this aberration.

Currently, the agenda is characterized predominantly by concerns about climate change, soaring energy and commodity prices, fair trade and ethical supply chains. There is also a swathe of regulatory interventions by governments around the world to mitigate the worst impacts of pollution caused by a global population of 6.7 billion people today (growing to 9 billion by 2050) seeking a higher material standard of living. But for all sorts of different reasons, biodiversity is not always seen as an equal priority.

'The Business of Biodiversity' nails down this mis-prioritization with splendid eloquence. Once business people come to see that biodiversity still represents 'the primary resource for all our business activities' the business case for embedding biodiversity right at the heart of corporate strategy grows stronger by the day. By the same token, the societal case for putting biodiversity 'at the top of the agenda' rather than treating it as an irritating afterthought becomes overwhelming.

Understanding this broader societal context is important. Hence many treatments of biodiversity take a narrow focus, homing in on particular species, habitats or biomes. Mark Everard keeps bringing the argument back to the way in which all our aspirations (for wealth, well-being, freedom, quality of life and so on) depend on protecting and enhancing the natural systems which underpin life on earth. Understanding this crucial differentiation between 'interconnectedness' (which simply reveals the extent of the relationships between ourselves and biodiversity) and 'interdependency' (which uncompromisingly compels us to acknowledge the importance of protecting biodiversity as a pre-condition to meeting our own needs and desires in the future) represents a crucial step-change in reframing the way we address 'biodiversity issues'.

Many of those 'issues' are thoughtfully unpacked along the way – in separate chapters on fisheries, forestry, farming, water, climate change and so on. Inevitably, a somewhat grim picture emerges, perhaps the most powerful articulation of which can still be found in the extraordinary Millennium EcoSystem Assessment, which brings together the work of more than 1500 scientists all around the world intent on providing politicians and business leaders with an overall summary of the combined impact of human endeavour on the natural world.

As Mark Everard points out, the speed at which governments managed to bury the report provides an all too telling reminder that they just haven't joined up the dots as yet, and really do continue to believe that it's possible to go on 'trading-off' the next chunk of biodiversity, indefinitely, without imperilling our very survival.

In 'Collapse', Jared Diamond's magisterial account of why so many civilizations engineered their own demise through the abuse of natural resources and ecosystems they depended on, he speculates as to what must have been going through the mind of that particular inhabitant of Easter Island as he cut down the final tree on the island, confirming his people's unavoidable extinction. Are we now nearing an equally momentous turning point, 'teetering as we are on the edge of disaster', in Mark's words?

We clearly still have time to pull back from that edge. This text is replete with case studies demonstrating how easy it is (relatively speaking!) to reconcile legitimate economic expansion with the prioritization of protecting biodiversity. Many of these case studies have been made possible by progressive companies all over the world re-defining corporate self-interest by explicitly putting a value on biodiversity – both in the short and the long term. And there's guidance galore for business leaders intent on bringing forward more detailed action plans to start making a real difference on the ground.

This is not altruism. It is hard-headed business acumen, based on the recognition that commercial success cannot be secured without attending to the basics that underpin each and every business. And the fact that it is also 'the right thing to do', from a moral and ethical perspective, massively strengthens the case for urgent, timely intervention by business and governments alike to start re-balancing the relationship between ourselves and our own living home.

Jonathon Porritt is Founder Director of Forum for the Future www. forumforthefuture.org.uk

Introduction

Biodiversity matters. We all know that: whether in our hearts or in our heads – or indeed both.

When we see images of the clear-felling or burning of vast areas of Amazonian forests, we need no one to tell us that this is neither right nor tenable in the long term. We all welcome fish in our rivers, birds in our garden and wild flowers in the hedgerows, and feel devastated if they are wantonly polluted, trapped, shot, or needlessly sprayed with pesticides.

When we hear sound bites about the extent of the human contribution to the degradation of marine fisheries and associated ecosystems or expansive tracts of forest, or to the spread of deserts, or the drying-up of rivers and wetlands, it is pretty clear to most people where current trends are likely to lead us.

We might tend our garden at home in a way that is more sympathetic to wildlife, or join a local Wildlife Trust. We might contribute to conservation initiatives in the workplace. Yet how powerful are we when it comes to understanding, and taking action about, the burden that our business activities place on biodiversity? Most of us, I suspect, feel pretty powerless, despite often deep-rooted concerns and intentions.

The purpose of this book is to help you take the first steps, both conceptual and practical, towards addressing the business of biodiversity. The arguments are not made in deeply technical language, nor does the book bombard you with excessive facts, figures and scientific references. There is a place for them, but this is not it.

This book is short – and deliberately so – to enable you to consume it in the space of a plane or longer train journey. However, the knowledge underpinning it is extensively researched, stemming from leading-edge thinking about the place of biodiversity in sustainable development and sustainable business. The topics are addressed in layman's language to help you get to grips with them yourself and begin to address them with colleagues at work.

Good luck with this task! After all, as you will appreciate from a quick read, the challenges facing biodiversity are real and pressing, as are the implications of depleted biodiversity for our human world, of which your business is a dependent part.

Part I Biodiversity basics

1.1   What is biodiversity?

The term 'biodiversity' is shorthand for biological diversity. Biodiversity is about the living world with which humanity not only co-exists but has evolved as a dependent part. Consequently, the natural world is neither remote nor disconnected from us, nor should we feel merely altruistic towards it. We and our activities are fully interdependent with the global ecosystems of which we are part, and whose fate we will inevitably share.

Understanding biodiversity requires recognising our interdependencies, and with them also our responsibilities. It can help inform us about redressing our behaviour to better ensure that our relationships with the world that supports our well-being and potential may continue in enduring harmony. This, obviously, includes the realisation of sustainable profit from business ventures that address the risks associated with their impacts on the natural world.

The emphasis of this book is on how biodiversity is best contextualised within business. It is not a thesis on the science of biodiversity for its own sake! However, to get us under way, we have to kick off this chapter with some notes on key biodiversity concepts. So here goes a brief, but not overly technical, overview.

1.1.1   Definitions of biodiversity

Biodiversity is not simply about 'lots of species'. Some habitats, such as naturally acidic lakes, many deserts and nutrient-poor streams, are particularly species-poor. However, the assemblage of species occurring naturally within them is characteristic. Different habitat types vary markedly in their natural attributes, and consequently support wholly dissimilar biological communities, with various species occurring in greater or lesser numbers. The unique assemblage of species of both microorganisms and larger plants and animals will vary from place to place. Even comparison of two or more habitats of a similar type often reveals a different balance of species, age structure and specific life stage within populations of single species (e.g. salmon fry in hill streams, salmon parr in the lower river, salmon smolts in the estuary and adult fish in the sea), as well as local genetic distinctiveness. This assemblage of life forms functions as an integrated and resilient whole – an ecosystem – adapted to the attributes of the unique habitat that has shaped it.

Biodiversity has been defined in many ways. A particularly useful definition is provided by Article 2 of the Convention on Biological Diversity (a global compact which will be described more fully later in this book):

> The variability among living organisms from all sources including, *inter alia,*
> terrestrial, marine, and other aquatic ecosystems and the ecological complexes

of which they are part; this includes diversity within species, between species and of ecosystems.

There are many more definitions of biodiversity in use today, several of which are included in the Appendix. However, we need not get too bogged down in the technical issues distinguishing them. Among the key common features is the fact that biodiversity is not just about numbers of species, but about appropriate and intact communities and the complex interactions between species, habitats, generations, genetic variability within species, climate and so forth. The term 'biodiversity' acknowledges that the natural world is a whole living system, and we are a part of it.

1.1.2 How biodiversity got there

The biological diversity that we live with today is bewildering in its abundance, breadth of different types of life forms and their intricate interdependencies. Surely no one can fail to be impressed by the range of imagery offered to us in nature programmes on the television or in colour magazines? From ornately plumed birds with equally elaborate display dances, the pole-to-pole migration of the arctic tern or the humpback whale or the dependence of one species of orchid on one particular species of fungus on its roots, there is a plethora of adaptations among the Earth's living organisms to impress and amaze. The refinement of detail and interdependence is stunning. For example, most of the many species of moth and butterfly caterpillars are evolved to eat only one or just a handful of plant species. In turn, these caterpillars have specific species of wasps that

live out their parasitic lives on them, and only them. Both of these relationships exhibit the most elegant co-evolution. As indeed do the many species of caterpillar that have come to resemble a particular type of bird dropping or dead leaf, or accumulate the toxins produced by a species of plant on which they feed and which then serve as a deterrent to their predators. The examples of beautiful symmetry and precision engineering within nature can go on and on. So too can the myriad elegant adaptations of species to habitats as diverse and bizarre as volcanic hot springs or inside aircraft fuel tanks, and from thermal vents on the deep ocean floor to the upper atmosphere. Some of the most ingenious relationships that have evolved between species are parasitic (where one species takes advantage of another) or symbiotic (where both parties gain from the relationship). Nature is rife with such perfect melding of form to function, elegantly attuned to the delicate set of balances and linkages on which it is reliant.

All of this biological wonder did not just get there by chance. Sorry if you are a Creationist, but the overwhelming evidence suggests that biodiversity did not just fall out the sky fully formed! Instead, the moulding of each life form and its unique life history have been shaped progressively by natural selection as a co-dependent part of a host of other life forms from the macro to the micro, from the immediate neighbour to the remote community and indeed all living things at the global scale. All are a product of close-grained interactive adaptation and evolution over billions of years.

We share our world today with unique assemblages of species honed by evolution since the origin of terrestrial life some 3.85 billion years ago. To state the obvious, that is a long time ago! It is also an awful lot of complexity, particularly when one considers that, in taking a breath, one is interacting both with local fluxes of water, air and suspended microorganisms, and with ecosystem-mediated air purification processes that operate at the whole-planet (biosphere) scale.

1.1.3   Biodiversity and human development

We *are* biodiversity – or at least a small part of it. As we've noted, the heritage of life on this small blue planet is, as far as we have been able to determine from the fossil and geological record, around 3.85 billion years. Reasonably complex life forms such as fish have inhabited the Earth for half a billion (500 million) years, first appearing in a geological period we now refer to as the Ordovician; this latter period represents only about one-eighth of the history of life on Earth. We, however, are recent introductions to the planet's fauna. Neanderthal Man, from the era we generally refer to as the Stone Age some 100,000 years ago, arrived only in the last quarter-of-a-ten-thousandth of the history of life on Earth. (Or, to put it in more graphic terms: if the whole of evolutionary history is represented as 1 day, humans appeared on the last second before midnight.)

Addressing the conservation of biodiversity is about how to meet the needs of a growing human population without compromising the variety and abundance of plants, animals, microorganisms and their habitats. Ignoring for the moment important moral issues, including our bequest to future generations, if we and our interactions degrade ecosystems then those ecosystems will in turn be less able to support us and our activities, including those of our businesses. And let's recall that we are not just talking about 'lots of species' like some giant museum, but about the diverse metabolic processes of all of the interacting living and non-living components of ecosystems. This includes the viability and productivity of soil, self-purification of water and air, production of other resources such as timber, fibre, food and pharmaceuticals, and the maintenance of climate and other life-support systems which are mediated by the interactions between micro- and macroscopic life forms.

It is arrogant to assume that current human knowledge embraces the entire depth and complexity of the finely meshed ecosystems that 'produce' and renew the resources that keep us healthy, wealthy and fulfilled. However, we do know

enough to begin to see that our collective activities are eroding the very life-giving systems that shaped us, sustain us now and may do so into the future.

So the importance of understanding biodiversity and finding a sustainable accommodation between us, our economic activities and the integrity of the natural world cannot be overstated. This is the case from the macroscopic and macroeconomic right down to the small-business scale, and we need to achieve this with far-from-perfect knowledge. However, we know enough to start the journey towards business that is in balance with biodiversity, and we can be assured that our knowledge will accrue and refine over time and can guide us to progressively more far-sighted decisions.

Although, as a scientist and naturalist, I will always feel the need to say more about the wonders of the natural world into which our own species appeared, and with which we and all of our activities interact inseparably, I have probably said enough to get us going in addressing the relationship of business with biodiversity. Let's press on, then, with understanding why biodiversity matters.

1.2   Why does biodiversity matter?

This is another broad topic, and one that we'll march through quickly to set the scene. As I've already mentioned briefly, biodiversity matters for a lot of different and sometimes seemingly disconnected reasons. These range from support for our basic survival through to providing a foundation for our economic activities and 'quality of life'. Biodiversity also has spiritual and religious resonances for different people, defines landscapes and natural regional character and is a focus of social value. All of this is ultimately germane to the linkage between biodiversity and business.

1.2.1   The heritage of life

How life originated on this planet 3.85 billion (or thereabouts) years ago will remain a mystery, though theories abound. The conditions at that time were inhospitable compared to today's standards.

Before life appeared, the major planets of the solar system (with the exception of Pluto) had coalesced from clouds of hot gases and dust circling the sun. Virtually all of the matter now on Earth was also there during its formation, the major difference being that it was then all mixed together in a largely homogeneous mass. A rocky core and mantle formed as the planet cooled, although much of the remaining matter was freely combined above them. Then, at an infinitesimal geological rate, heavier atoms began to be precipitated and immobilised, and water could form on a more solid surface. After billions of years, a level of purity adequate for the appearance of primordial life had been achieved.

Life changed this world, processing the chemical soup of the early seas and accelerating the process of biomineralisation (locking substances away into rocks as they formed in the Earth's crust). One of the most dramatic events in evolution took place around 2.5 billion years ago, when life had already been in existence for nearly 1.5 billion years. The first photosynthetic activity took place, liberating oxygen in its highly reactive molecular form into the atmosphere as green cells captured the energy of sunlight to build sugars and other complex molecules. This would have been a catastrophic event for most life forms of the early Earth, which had evolved in oxygen-free conditions.

Yet, from the germ of those organisms that survived, the evolution of life was accelerating. Photosynthesis led to the building of a dense, oxidising atmosphere. These conditions allowed the formation of a stronger shield against cosmic radiation and supported meteorological patterns and cycles, and the more rapid purification of the upper parts of the planet through increasing rates of both the chemical and biological processes. Progressively, more of the heavier chemicals

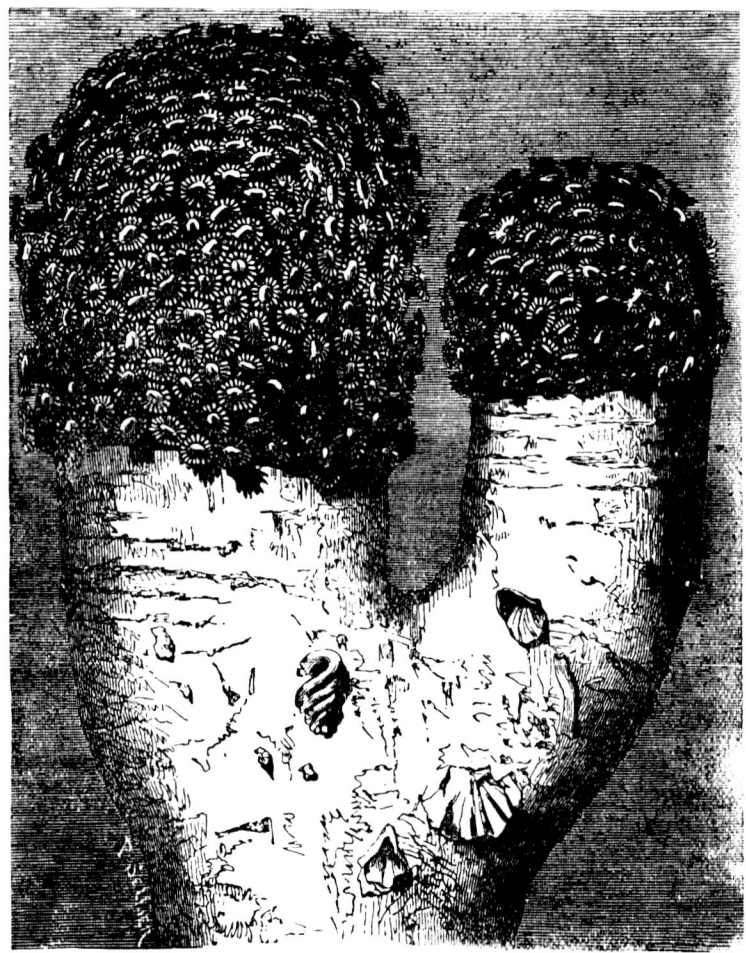

were precipitated and compressed into inert forms in the Earth's crust. All of this took the brakes off the pace of evolution by creating more favourable environmental conditions, containing less of what we might term today as polluting substances and therefore allowing scope for more advanced forms of life formerly inhibited by the early planet's toxicity.

Through the new and strengthening cycles of matter and energy processed by diversifying life forms, substances such as heavy metals, radioactive nuclides, chemical nutrients and carbon were progressively eliminated from unconstrained circulation, and locked away in the lithosphere (the largely inert rocky part of the planet).

The process of evolution and atmospheric purification has continued through subsequent billions of years. Many evolutionary blind alleys were embarked on. Today, we are in the flow of continuing diversification, the genetic blueprints of all living forms representing the legacy of billions of years of natural selection, adaptation to changing environments and resilience in the face of variability in climate, flood and other local conditions.

All of life, all of its processes, functions and cycles, resilience and adaptations, are the fruits of those billions of years. All of this, and many more aspects besides which we probably do not yet suspect, far less understand or recognise, comprises contemporary biodiversity.

Biodiversity has been time-hewn for half of the lifetime of the planet itself. It is the fertile soil of genetic tools and adaptations from which our species germinated in recent times and is, whether directly or through the processes it performs, a primary resource for all of our business activities.

1.2.2   Engines of creation

The cycles and processes that life performed as it evolved, keeping energy and matter in circulation and maintaining the atmosphere as a sustainable whole, remain with us today. Through fantastically intricate pathways, many seemingly redundant or barely understood and many more doubtless wholly unsuspected, the web of life keeps matter and energy in process from the upper atmosphere, across the land, through fresh and salt waters, to the deepest oceans and pores in rocks. It is these biological functions, intimately tuned and balanced by evolution, that keep the cycles of nature going and build the whole integrated life system that has formed symbiotically on this planet.

Like a super-organism, the totality of planetary life forms an integrated whole that maintains the atmosphere to maximise its own prospects of survival. This cohesive, whole-planet scale of the ecosystem is technically known as the biosphere, but is referred to more commonly by the name 'Gaia'. (Some use the term 'Gaia' with religious, 'Earth Mother' overtones, although the atmospheric scientist James Lovelock who originated the term certainly does not.) If this way of expressing life seems near-spiritual, it is not intended to be so. It is just that comprehending the sheer scale, intricacy and intimacy of interconnection of life on this planet, and its capacity to adapt to changing conditions and new niches, is overwhelming for our parochial human psyche. When we appreciate that we are but a recently evolved additional part of the greater whole of the biosphere, perhaps we should not be so surprised to find ourselves humbled by its scale!

For the purposes of this book, we should at least find pause to be impressed. Every life form on this Earth evolved integrally with this great system, and depends on it totally for its continuance, realisation of potential and reintegration after death.

1.2.3   Our part to play

Humanity is no exception to this rule of interconnectedness with the web of life. The fact that we breathe, drink, eat and excrete is itself full testimony to our complete inability, akin to all planetary life forms, to do anything that does not deeply interact with the rest of the Earth's complex ecosystems, which have been honed by nearly four billion years of evolution. Species, genetic and habitat diversity embody the 'intelligence' of these eons of evolution, with all species fully dependent on the integrity of the whole. Naturally, we too are utterly dependent on the Earth's diverse life forms – the 'cogs' of the natural cycles of planet Earth. We depend on our home planet's ecosystems totally for our health and potential, harness them for our wealth-creation activities, and need them to fulfil our individual and collective potential. It is that basic and that essential.

As social creatures, our civilisations depend on the functions of ecosystems within the web of biodiversity, from the provision of fresh air and water to productive soils for food and materials, and the dilution, breakdown and reassimilation of wastes. We arose from, remain within, and will ever be part of the ecosystems of this Earth, including all we eat, drink, breathe, utilise as economic resources, convert into products, throw away or rely on to absorb our wastes. In short, we and our activities, including our business ventures, are part of the biodiversity of our home planet.

1.2.4 A common destiny

It follows that, if humanity is an integral element of the biodiversity of this planet, we and it share a common destiny. If the global ecosystem thrives, it can better support our needs. If river catchments function effectively, they can provide us with water, fish and building materials such as rushes and timber, and sustain a

host of local needs. But if our activities, individually, corporately or collectively, erode the natural world and its supportive capacities, we erode the very systems that sustain us now and into the longer term.

This observation applies as much to our physical well-being as to our economic progress. Beneath the often most pressing daily matters of profitability and wealth creation, our economic activities are merely means for serving society's needs, all of which ultimately have a biological basis.

1.2.5 An evolving concept

The scientific reality is that we remain inexorably part of nature, despite protestations to the contrary by some religious doctrines and philosophies. Yet society acts in many ways as if this were not the case, drawing a Cartesian division of mind from body, and of human from nature. It is at our peril that we continue to act as if we are anything other than integrated, as we will inevitably suffer if we continue to take for granted the many functions of the natural world that make survival and economic activities possible. When we degrade global ecosystems, either by massive disruption or via the insidious, cumulative impacts of 'death by a thousand cuts', we inevitably reduce the potential for them to allow us to lead fulfilled lives, including our economic prospects.

We are learning that biodiversity matters for many reasons: for its own sake, because we are part of it, because it came before us and will endure when the age of humans is gone and because it makes our lives possible, profitable and enjoyable. And, embedding as it does the natural wisdom distilled by billions of years of evolution, it probably matters for a whole lot more reasons besides; reasons that we have yet even to suspect.

1.3 Biodiversity and business

For business, the focus of this book, biodiversity is the mother lode of resources for society's activities. It sets the limits for resource availability and absorption of wastes, and it is the needs and wants of the biological entities that we call 'customers' that determine the products and services that business supplies.

1.3.1 The capitalist ecology

For better or worse, the capitalist system is pervasive across developed and most industrialising societies. Once, humanity met its needs by itinerant harvesting and hunting, progressing at some stage thereafter (probably around 8,000–10,000 years ago) to a phase of settled agriculture that anticipated future yield from crops sown and animals reared. The emergence of trading mechanisms within society, together with development of language, writing and other communication mechanisms, marked the point at which true civilisation commenced. These mechanisms offered a means for the differentiation of the tasks that keep society going, enabling it not merely to support itself but to develop beyond basic biological needs.

Since that time, the human invention of money has held a variety of meanings for people. We have no space here to discuss this fascinating history, other than to note that money is a means to an end, that end being society's exchanging of things to service its wants and needs. And, since these wants and needs depend on natural resources, the economy is merely a mechanism by which human society continues to metabolise as part of the supportive cycles of nature. Money may appear in many cases to separate us from nature, but in fact it is a means to join us to the goods and services that biodiversity has to offer. After all, the very words 'ecology' and 'economy' stem from the same linguistic source: the Greek word *oikos*, meaning 'house'.

Hence business is a device within a capitalist economy that converts basic resources into useful services and products supporting human needs and wants through a wealth-creating process. In theory, if all costs and benefits are factored into our trading mechanisms, we have a system attuned to balancing society's needs with the capacity of nature to provide for them.

Problems can arise where we break that link, ignoring natural limits and thereby becoming a parasitic (exploitative with no return) rather than a symbiotic (achieving mutual benefit) element of the world that supports us. Then, the human-made economy would, like cancer cells, develop aggressive self-serving growth patterns disconnected from its purpose of serving society. The consequences, like cancer, would be destruction of the host and, just as inevitably, the eventual death of the cancer cells too. There are indeed many parallels between the progression of cancer cells and the relatively unconstrained economic system of contemporary consumerist society, which is progressively influencing more of the world to generate short-term profit with little concern for the long-term consequences for global biodiversity and integrity. Indeed, it seems today that much of society's activities service, rather than are served by, its economic creation, eroding the life-support mechanisms of the world for short-term benefit. (I will return to this analogy later in the book.)

Let us keep in mind the concept that business is the means to the end of serving society's needs. Although business was never designed with the intention of being the main driver of current massive biodiversity loss and other environmental problems, it demonstrably has that power. Also, it can use this power in more foresighted ways to become one of the key factors in curing society's inherited systemic problems. As we will see in the pages of this book, there is real business advantage in restoring commercial activities so that they operate within the supportive capacities of biodiversity. Not least among the arguments for doing so is that, if we destroy the host (human society and the ecosystems that support economic activities), we ultimately destroy our resources, our markets and ourselves.

In extremis, this will be a literal conclusion. However in the meantime we have often experienced insidious effects indirectly, including through resource scarcities, spiralling costs, difficulties in securing development planning, customer and societal acceptability, corporate reputation and so forth.

1.3.2   Goods and services

As society becomes increasingly concerned about and confronted by its impacts on the natural world's systems and biodiversity, and aware of the need for a sustainable basis for this relationship to assure long-term well-being and prosperity, biodiversity and other attributes of the Earth's ecosystem will increasingly set parameters shaping the future of business.

The two key ecosystem concepts of 'goods' and 'services' (outlined below), in addition to the moral implications of society's relationship with biodiversity, have both direct and indirect relevance to business. These ecosystem concepts of 'goods' and 'services' translate reasonably directly to the mainstream business language of 'products' and 'services', the latter pair being the tradable items stemming from utilisation of the former in the process of wealth creation. And, although wealth creation serves a component of human development, the wider implications of business activities on biodiversity and its various functions and public benefits necessarily form part of the 'licence to operate' that society grants to businesses.

Bringing biodiversity considerations into far-sighted strategies and decisions is, then, central to shrewd business judgement, germane to both the core business model as well as the broader relationship with the diverse stakeholders on which sound and sustainable profit depends.

1.3.3   Goods and products

The stocks of organisms, and of biologically derived (e.g. fibre in clothing or paper) or biologically mediated substances (such as mineral deposits), constitute tradable goods and raw materials for manufacture of further products. Timber and a host of other forest products, from harvested nuts and fruits through to paper pulp, are pertinent examples of ecosystem 'goods' that translate directly into tradable products. So, too, are those fish stocks exploited for human food, sport, industrial processes or agricultural inputs. Thatch for roofing, most frequently derived from beds of the common reed *Phragmites australis*, and charcoal, produced from a range of sources of wood, are further instances of trade based directly on the harvest of standing crops of biota.

If only nature was still so abundant that we could live our lives through the relatively passive harvesting of nature's excess! However, the demands of a burgeoning human population on dwindling biospheric resources means that stewardship of natural stocks or managed production through enhanced productivity, typically through intensive farming, is the norm today. And, although managed farmlands are one step removed from direct harvesting of tradable goods, the agricultural process is merely a managed intensification using natural productive systems such as soil, water, nutrients and selected strains of living organisms.

Natural stocks of gravel, chalk, fossil fuels and other mineral deposits represent fossil concentrations of useful 'goods' produced by the historic action of natural processes operating across the ecosystem. Biodiversity has had more than an incidental hand in their concentration into useful deposits and forms.

All of these examples illustrate how the ecological concept of 'goods' can be translated directly into the concept of market goods or products. However, we know from experience how this process can lead us to override natural limits to productive and regenerative capacity, which may in turn result in ecosystem breakdown. Sadly, the economic system that we have inherited from our Industrial Revolution forebears externalises, or omits as a basic consideration, many of these biodiversity concerns. (We will explore this in far grater detail later in the book.) What may make sense from a narrowly economic angle – a path that businesses may feel obliged to follow in fulfilling their obligation to return maximum value to shareholders – is not automatically geared to the protection of the ecological 'goods' that support the enterprise, nor the unwelcome prospects for long-term human well-being as a consequence of their degradation.

1.3.4   Services

Both ecosystems and businesses produce 'services', and both provide value to people. For example, the water cycle of this planet is founded on a familiar pattern of evaporation and subsequent precipitation of pure water, which then falls on catchments (watersheds) to support ecosystems that include human sustenance and economic activities. In their natural state, catchments acted like giant 'sponges', absorbing and storing water, purifying it through various ecological processes, and releasing it as a smoothed, steady flow throughout the year. Aside from supporting diverse life forms, the ecosystem services of water purification and hydrological buffering (the process of water storage and smoothing of flows) have clear economic value. They provide the raw resource for the water industry, and other industrial sectors that utilise water in their processes, to service society's demand for water and removal of water-borne waste. They also support other businesses by enabling abstraction for irrigation and industrial purposes, wet fencing of stock, and water-based transport, sport and defence. Furthermore, steady flows of pure water also provide a medium for the dilution, absorption and metabolism of much of society's wastes. Take away the services of the water cycle – or diminish them by short-sighted land drainage, unsympathetic management, overburdening with waste, damage to habitat through dredging or insensitive logging, farming or urban development – and the capacity of catchments to support these myriad human needs and economic activities is also degraded.

The same analysis of ecosystem services could be made for the complex environmental processes that purify the air and moderate the climate at both local

and biospheric scales, and for the capacity of fertile soils to support human needs ranging from the production of food to energy crops, fibre, dyes, chemical feed-stock and pharmaceuticals.

As an exact parallel to the aquatic situation, it is equally clear that the massive levels of soil erosion seen both in the UK and worldwide are actively undermining the 'service' of soil productivity, which will automatically limit future options for human well-being and economic activity.

1.3.5   The ethical dimension

In addition to this translation of ecosystem goods and services into economic products and services, the protection of biodiversity has a strong ethical dimension. This may not be widely recognised in the ethos of traditional capitalist business. Indeed, it may be seen as a marginal and ill-defined concept with flaky spiritual connotations. However, this aspect of humanity's relationship with biodiversity is as hard-edged and as business-relevant as any other.

From the moral or ethical dimensions, leaving aside more contentious spiritual or religious interpretations, biodiversity is seen as the inheritance of current generations, which are therefore its stewards, charged with ensuring it is passed on undiminished as a bequest to future generations. However, this commonly articulated view is itself anthropocentric, disguising the fact that it is we who are a mere recent addition to the catalogue of biodiversity of this planet, and one with no automatic right to continued existence through the processes of natural selection.

Yes, we owe it to future generations to pass on an undiminished, and ideally a restored, wealth of biodiversity such that the biosphere can provide the goods and services that may allow them to meet their needs and realise their potential. However, biodiversity has an existence value too: its own intrinsic worth, which is completely independent of any utility we may find in it. It is a precious residue of billions of years of evolution, out of which we ourselves have so recently arrived. It is biodiversity that has the venerable status in this world, and we have no right to wantonly or carelessly destroy it for short-term gain. Indeed, given our tight interdependence with biodiversity, we can only compromise our own prospects if we do this.

It is this instinct, often deeply hidden but nevertheless encoded in our evolutionary psychology, that is triggered when there is public outcry over the more devastating business impacts on the natural world: dead fish, denuded or felled forests, fishless seas, silted or drained waterways, suffering stock and so forth. Businesses offend this basic human instinct at great peril to their reputation.

1.3.6   Ecosystem services

Today, the diverse beneficial goods and services provided by ecosystem are now amalgamated within the term 'ecosystem services', driven largely through a redefinition within the Millennium Ecosystem Assessment (MA).

The MA comprises the most authoritative assessment of the status of global ecosystems, the 'overshoot' of resource exploitation by humanity over natural regeneration rates, and the implications for long-term human well-being and security. The MA was initiated by former UN Director General Kofi Annan in order to assess humanity's impact on the natural world, implications for human progress and options for future pathways for sustainable development. The first MA reports were published in 2004/2005 [1, 2], reflecting the work of approximately 1,300 scientists worldwide and supported by the consensus of many thousands more. They demonstrated unambiguously the rapid decline in all the major ecosystems of the Earth, together with their capacity to support human industry, food and potential into the future. The conflict between booming population and declining biospheric carrying capacity is, without question, the greatest challenge facing our collective future. The MA also refined understanding of the many beneficial ecosystem services on which all human activities ultimately depend, positing an 'ecosystems approach' as a basis for more foresighted, ecosystem-focused development to shape a more sustainable future by decision-makers at all scales from the global to the national and indeed business scales.

The fundamental principle of the 'ecosystem services' approach is recognition of the ecological basis of the primary resources that make life possible, profitable and fulfilling for human society. Ultimately, all people depend on the productivity of natural systems, no matter how far removed from the basic resources we may be. Today, all global ecosystems are recognised as providers of often overlooked yet essential and irreplaceable services to humanity.

Although we will not go into the finer details of the Millennium Ecosystem Assessment's 'ecosystems approach' in this book, nor the various tools subsequently produced for their application in policy development and economic assessment, it is worth recording that it classifies ecosystem services into four broad categories – provisioning services, regulatory services, cultural services and supporting services – breaking these further into generic sets of services as detailed below.

Provisioning services:

- Fresh water
- Food (e.g. crops, fruit, fish, etc.)
- Fibre and fuel (e.g. timber, wool, etc.)
- Genetic resources (used for crop/stock breeding and biotechnology)
- Biochemicals, natural medicines, pharmaceuticals
- Ornamental resources (e.g. shells, flowers, etc.)

Regulatory services:

- Air quality regulation
- Climate regulation (local temperature/precipitation, GHG sequestration, etc.)
- Water regulation (timing and scale of run-off, flooding, etc.)
- Natural hazard regulation (i.e. storm protection)

- Pest regulation
- Disease regulation
- Erosion regulation
- Water purification and waste treatment
- Pollination

Cultural services:

- Cultural heritage
- Recreation and tourism
- Aesthetic value
- Spiritual and religious values
- Inspiration of art, folklore, architecture, etc.
- Social relations (e.g. fishing, grazing or cropping communities)

Supporting services:

- Soil formation
- Primary production
- Nutrient cycling
- Water recycling
- Photosynthesis (production of atmospheric oxygen)
- Provision of habitat

This, however, is a modern take on our relationship with nature. It hardly reflects the legacy of our Industrial Revolution past which still substantially shapes the governance and financial compulsions and inherited assumptions under which business operates. We have to examine these next if we are to understand the challenges and opportunities for business in taking a more foresighted approach to its dependence on biodiversity.

1.3.7   The business response

From the time of the Industrial Revolution through to modern environmental consciousness, business was generally simply unaware of its impacts on a natural world that was assumed to be limitless. We can find refuge in no such blindness today. Indeed, for a number of years now, many leading businesses have sought to take a more proactive and responsible attitude towards biodiversity. However, it is a sea change in the practice of mainstream business that we are looking for if we genuinely want to reverse our unfortunate habit of destroying the natural world in the name of progress.

The MA has obviously further focused international attention on the challenge of addressing our interdependence with the natural world, and the need to innovate new industrial and other cultural habits if we are to safeguard the primary resources that underwrite our common future. There have been many subsidiary supranational and national responses. For example, EU-wide initiatives call for stemming the tide of loss of biodiversity whilst the UK government's Environment, Transport and Regional Affairs (EFRA) Select Committee drew attention to the need for business to engage more fully in the conservation of biodiversity in its report *UK Biodiversity*, published in December 2000 [3]. Allied to this, there have been a number of government challenges to businesses to produce social and environmental reports, develop coherent business cases to manage biodiversity issues, integrate biodiversity considerations into existing company management systems, and develop tools and indicators to help with progress. Often, however, it is simply a case of lacking a practical purchase on the issues, and lacking the pragmatic measures to make a start on a problem that can initially appear intractable. This is where this simple and short guide comes to your aid.

1.3.8   Trading on biodiversity

The net conclusion of this rapid overview is simply that commerce is in no way different to any other area of human endeavour, and that it too interacts fully with the ecosystems of our home planet. Wealth-creating business activities that convert ecological resources into products and services to support social well-being must be conceived as an integrated whole. Sustainable development rests on the simultaneous fulfilment of all three dimensions – economic, environmental and social – and this principle applies equally at the biospheric, national, regional, local and corporate scales.

The challenge is to translate this high 'motherhood and apple pie' set of principles into a pragmatic basis for factoring biodiversity thinking into business decision-making. In the following chapter, we'll explore some of the ways in which the benefits and burdens of our use of biodiversity fall on different communities, or other sectors of society. Then we will explore wider links between business and biodiversity, and the practical ways in which they may be integrated into mainstream decision-making.

References

[1] Millennium Ecosystem Assessment. (available at: www.maweb.org).
[2] Millennium Ecosystem Assessment. *Ecosystems and Human Well-Being*, Island Press, 2005.
[3] Environment, Transport and Regional Affairs Select Committee, *UK Biodiversity*, The Stationery Office: London, 2000.

1.4 Benefits and burdens

The theory of market economics is that everything has a recognised value that is factored into trade. This appears nice in theory, but is disastrously flawed in practice. In the 'real' world, many of the 'real' values of all resources, both the ones that can be monetised and the ones that cannot, are simply excluded from market prices. This market 'externalisation' of biodiversity and other core resources has serious implications for both biodiversity and human equity, and lies close to the root of many of the systemic environmental problems causing havoc in our world today.

1.4.1 The perfect market

The perfect market economy model is an invention of economists that rests on a suite of assumptions. These include that there are no 'distortions' (such as taxes), that all property is privately held, that all human utility and enjoyment can be measured in financial terms, that all assets are priced and traded in markets, and that there are no measurement problems or transaction costs. Furthermore, there is an implicit assumption that there is no crime or war (non-market costs), that litigation is cost-free, that there is perfect competition, and that all information is known and frames people's choices and behaviours. Rationality is also assumed, wherein all players in the market make decisions based on logical economic reasoning, a difficult assumption to accept when one looks at the impact of lifestyle values and choices, fashion and marketing hype on corporate performance. What is so very rational about money spent mopping up the harm from pollution or rampant crime being counted as a positive contribution to the economy?

Critically for biodiversity, the perfect market model overlooks externalities such as pollution or exploitation of biological resources beyond natural regenerative capacity. It also overlooks publicly held goods, such as oceans, the atmosphere, mountains, deserts, birds and other such 'commons', assuming that market price can reflect their inherent value as transparently as privately held resources (see Section 1.4.3). Furthermore, and rather worryingly, there is an assumption that all human capital can be sold, which brings with it the inherent conclusion that slavery must still be rational and legal. From a developed/developing world perspective, this model can also overlook as irrelevant those people excluded from the dominant global economic model but nonetheless suffering directly from exclusion from resources such as water and food. Also, the economic model is generally used to argue that what is good for consumers and producers is also good for society as a whole, despite overwhelming evidence to the contrary.

Another assumption at odds with the 'real world' is that the economy has no institutions, such as monetary, legal or government systems, or else that these institutions exist but can be ignored since they work perfectly at no cost. A further assumption is that there are no changes in society that may affect its preferences. Also, it is necessary in the perfect economic model for producers and consumers to freely enter and leave the market, with competition (even international competition) normalising profits and balancing the needs of consumers in a growth area of the market. Is there room for copyright, intellectual property and trademarks in this perfect world? Is there room for nomadic and traditional lifestyles enjoyed by people subsisting on the supportive capacities of ecosystems yet not participating, whether by choice or active exclusion, in the capitalist economy?

The perfect market economy model is an interesting thought experiment, assuming away all imperfections to describe what would be perhaps the ideal conditions for capitalist marketplaces to function. However, even the most cursory scan of the assumptions within the model reveals myriad 'imperfections' in the real world. Yet, despite near-ubiquitous acknowledgement of its failures, the market model shapes so much of our cultural and institutional governance. Is it surprising that such externalised 'details' as the very life-support and productive systems that keep society and business going are withering away under the unsympathetic policies of a 'perfect' human conception of the world?

1.4.2   Present imperfect

Welcome to the real world! Ours is a world where, for all the product labelling we see, the public does not immediately receive full access to information about environmental, social and financial impacts. So time-poor are we that, even if the information were known and provided, and we were sufficiently educated to interpret it, we would probably have no time to read it. Neither do we live in a world of perfect competition where sellers are perfectly matched with buyers. Instead, buyers and sellers are able to influence prices – a condition not allowed in the perfect model – whilst many in global society do not even play the game of capitalism.

Furthermore, and contrary to the song in the musical *Cabaret*, money does not make the world go round! Money is a human construct, invented to enable trading, as society became settled and people engaged in different activities to support the collaborative whole. The hierarchy is: biodiversity and its myriad processes support human needs and allow for economic activities; these activities are geared to supporting society's aspirations and money is a recent invention to provide the capacity for equitable trade across society. The perfect market, as assumed by many political institutions and regimes (*It's the economy, stupid!*), asserts the primacy of the economy. The reality is somewhat different, stemming from our inescapable biophysical dependence on the natural world, which is inadequately factored into economic theory – a relationship that is inverted in the economic and corporate governance systems we have inherited.

Like Dr Frankenstein, we have created a creature intended to serve the purpose of good only to find that its self-determining steps have ultimately also produced the opposite effect. Money becomes the goal, competition with trading partner businesses or nations the prize, and social needs and impacts on biodiversity a troublesome by-product. This is an odd take on the concept of 'perfect' when we consider the massive degradation of natural supportive ecosystems around the world, with their inevitable adverse consequences for humanity!

Implicit within the 'perfect economic model' is that all goods are tradable. Therefore, if we are short of food, a basic biological necessity, we can buy it from somewhere. If climate change exacerbates flooding and storm damage, we can pay to repair it. Yet, if the world at large is short of food, water or clean air, threatened by massive climate change, or depleted in natural resources including the capacity to absorb society's multiple waste streams, who exactly can we pay to fix it? Where will they procure the goods? And, in any case, can we afford the price? In the real world, we know that we cannot eat, breathe or drink money, and that certain things have a heritage or even sentimental value, or deserve protection from our activities just because 'they are worth it'. Are conservation policies, World Heritage Sites, family bonds, aesthetic values, societal cohesion and other non-market goods really worthless, or is the economic system itself in need of an overhaul?

1.4.3 'The tragedy of the commons'

Many things exist outside of the perfect market. In fact, this includes virtually everything that is of fundamental importance to the continuing integrity and stability of the ecosystems that support human well-being and aspirations. The term 'the tragedy of the commons' is now in common use to describe the consequences of over-exploitation of commonly shared goods not under individual ownership.

The phrase itself derives from a parable published in 1833 by William Forster Lloyd, Drummond Professor at Oxford and Fellow of the Royal Society, in *Two Lectures on the Checks to Population*. This was subsequently elaborated and brought to wider public attention by Garrett Hardin in 1968 in an essay in the journal *Science* titled 'the tragedy of the commons'. The tale narrates how unrestricted access to a common resource, without common agreements, ultimately dooms the resource due to the fact that the benefits and burdens are shared inequitably. In the cases of overgrazed common meadowland, international marine fisheries or any of a host of other environmental 'commons', individual (or national) benefits stemming from exploitation create an incentive for an increased 'take' since the costs of exploitation are distributed between all those exploiting the resource. Hardin extended the narrative to focus more generally on the use of common biospheric resources such as the atmosphere and oceans, as well as draw attention to the 'negative commons' of pollution. Individuals then compete to maximise their take from the common before others beat them to it. No one owns the common good, which inevitably withers away under aggressive competition since all overlook its natural regeneration rate. Hardin also highlighted that energy inputs to this planet are finite, and that the 'tragedy of the commons' metaphor may be extrapolated to conclude that there is no foreseeable technical solution to increasing both human populations and their standard of living on a finite planet. While this may or may not be true, it is certain that the biosphere of this planet is a common that we are exploiting well beyond the finite limits of its capacity to regenerate. No one owns it, no one pays for it, but we are all desperate to take a piece before others beat us to it.

If we need practical illustrations of this phenomenon today, we need look no further than the collapse of forest ecosystems, marine fisheries, aquifers or the structure and quality of soils observed worldwide. (We will look at human impacts on some of these attributes of biodiversity later.)

In the absence of traditional collective stewardship arrangements ensuring the long-term well-being of a common resource – for instance, traditional nomadic Maasai management of grazing lands prior to European enclosure of East African savannah lands – concerted action by government and consortia of common interests is necessary to resolve 'the tragedy of the commons'. The direct action of self-interested individuals cannot alone promote the public good because the market power opposing such minority interest is too great. To resolve the conflict, we require management solutions that may include strong conservation and sustainable yield legislation with associated permitting, privatisation of certain resources with associated checks and balances, application of the 'polluter pays principle', and a strong regulatory climate. We also have to make sure that these arrangements are informed by robust science and actually work, and that international agreements do not simply descend into national bartering of exploitation rights such as we commonly see in fishery agreements worldwide. (Again, more details of that later.)

Although we may in some cases understand the problems affecting these 'commons', the evidence of forest loss, fishery collapse, groundwater over-exploitation – and most environmental 'commons' besides – is that we are as yet highly inefficient in implementing effective political responses to head off degradation of biodiversity. As long as commonly owned goods and wider benefits to the 'common good' remain externalised from the market, and in the absence of effective stewardship legislation and incentives (such concepts being alien to the 'perfect market' model), nature and its capacities to sustain human well-being and business success will inevitably decline. The need for a long-term accommodation between biodiversity and business is pressing, along with normalising legislation and other measures to ensure that such 'commons' are managed equitably and sustainably with no excessive burdens placed on businesses seeking to be more responsible.

Many biodiversity-related resources are commons, and therefore usually externalised from markets. It is therefore first necessary to begin to account for some of the environmental dependencies of business and then to seek means by which their sustainable exploitation can be assured. There can be perversities in the inequity of costs and benefits from exploitation of commons. For example, in a river catchment, unwise land use such as tillage of catchment-critical habitat, insensitive ploughing promoting soil erosion, use of fertilisers or excessive abstraction near upland water sources can provide economic return for landowners and managers. However, stifling the various ecosystem functions provided by these headwater ecosystems then reduces the quality of water, reliability of flow, groundwater storage, recruitment of fish fry and invertebrates for lowland fisheries, and disturbs silt flows and fluxes of nutrients and other aspects of the 'common' of the whole river system such as natural beauty and bird life. Not only that, but money collected from taxpayers in the lower catchment is often used to provide support payments for the destructive land use actions by farmers at the top of the catchment! This, of course, is just one of many examples of an ecosystem in which society not only rewards degradation of biodiversity and environmental integrity but also reinforces it with further economic perversities.

Clearly, we are in need of a strong policy environment that places biodiversity and its support functions close to the top of the agenda, rather than as an afterthought to be conserved locally using a small fraction of the revenues generated through business processes that contribute to its ongoing and widespread destruction.

1.4.4   Intergenerational equity

It is not, of course, just equitable access to resources across today's society that matters. There is an overriding moral issue associated with our legacy to the future, including our own offspring. The concept of 'intergenerational equity' encompasses the belief that different generations should be treated in similar ways, and should have similar opportunities.

Human well-being, economic opportunity and 'quality of life' all stem from biodiversity resources, which provide people with 'goods' and 'services'. This includes the consumption habits of today's generations across the wide age range within society (i.e. 'generational equity') as well as the legacy that this will leave for generations as yet unborn ('intergenerational equity'). To this extent, the future is a 'common' that we exploit at net cost to its other potential users. Market economics, if left unconstrained, would sadly lead us to maximise profit today: hang the consequences for tomorrow!

Intergenerational equity is a value-based concept addressing the rights of future generations, and is a component of any serious conception of sustainability. It therefore extends the scope of social justice into the future, and also of the wise stewardship of biodiversity and other natural resources. According to the moral of intergenerational equity, each generation has the right to inherit the same diversity in natural and cultural resources enjoyed by previous generations. If we break this duty, we leave them with impoverished ecosystem services with which to meet their needs and aspirations. We therefore owe them a set of economic and governance systems that are more far-sighted and stable than those we inherited. However, it is only by taking account of and protecting biodiversity, or better still restoring it, that we can protect the needs of generations to come by passing on the basic resource that underpins their needs and aspirations. Any other conception of delivering intergenerational equity – for example by maximising profit today to hand them a strong economy tomorrow – risks missing the most fundamental point of sustainability and ignores the resource base that sustains it.

Societal valuation of biodiversity then becomes a central plank in its legacy for future generations. On that account, we are currently seriously in deficit when we look at the degradation of natural resources across the globe. We have work to do to bring biodiversity into the centre of our thinking, decision-making and actions.

1.4.5   The absolute value of biodiversity

Economic valuation of the natural world is now pervasive in the socio-political deliberative processes of the West. The primary determinant of economic valuation is human utility, as it is this aspect that can be related to monetary value and 'market price'.

While economic exploitation of natural resources is as old as human civilisation, the field of environmental economics is relatively more recent. It was substantially accelerated by the work of Partha Sarathi Dasgupta throughout the 1970s on the economic theory of depletable resources, green accounts and the importance of essential ecological services and environmental resources for poor countries. Dasgupta, a professor of economics at the London School of Economics, restored an ethical element to economics by highlighting the direct link between social well-being and the quality of the natural environment that provides for human needs [1]. The field of environmental economics flourished thereafter, particularly following the 1992 'Earth Summit' on sustainable development at Rio de Janeiro. Many methods have been used to try to monetise the functions of nature in terms of the goods and services exploited by humanity. For some species and habitats, public surveys have been used to derive 'willingness to pay' values, or else the 'contingent valuation' of how they add to house prices or other assets have been deduced. A common approach is calculation of replacement costs.

An important milestone was set in the history of environmental economics with the publication of a paper by Bob Costanza *et al.* [2]. Costanza *et al.* conservatively estimated the value of all of the Earth's ecosystem services, on the basis of replacement costs at current market rates, at $33 trillion a year at least. This was close to the world's total gross domestic product at the time. This paper still remains the best-known, albeit speculative, attempt to ascribe monetary values to the ecosystem services from which society benefits on a global scale. Even if it is a couple of orders of magnitude out in its estimates, the figure is hardly insignificant! However, the trouble even with this conservative 'ballpark' estimation of nature's collective worth is that it seeks to place monetary values on essentially non-monetary resources.

If biodiversity were purely a human-made concept, it would be possible to ascribe to it a human-produced economic value. However, it is not. Biodiversity instead is a genetic heritage born of billions of years of evolution, including myriad interdependent life forms, a fantastically efficient, adaptive, resilient and almost infinite network of pathways for processing chemicals and energy, and an irreplaceable life-support system that provides for human potential. It is arrogant in the extreme to think that we can ascribe it a human-made economic value.

A 'deep green' conception of biodiversity starts with an assertion that the natural heritage of this planet is of inestimable value because the survival of all its components, including the human species and its aspirations, relies on its integrity. This sense of an absolute value of nature is missing from the economic systems of the developed world. And here lies another flawed assumption: that all may be traded or otherwise 'developed' without loss to human capacity in the future.

Today, the Millennium Ecosystem Assessment's reconceptualisation of ecosystem services provides a novel basis for the more inclusive accounting of nature's supportive mechanisms. And, as we will see later in this book, they have provided a useful basis for better accounting for the value derived from ecosystems as well as the marginal benefits and costs of different development options. However, the central issue of relevance here is that we have to learn to see ourselves not from the Industrial Revolution paradigm as the users of a limitless world, but as interdependent elements of complex ecosystems that we harm to our own long-term and often immediate detriment. And we need to make this shift practical in the decisions we make in business if we are to contribute to a more sustainable society in ways that also offer us better protection from the economic consequences of short-sighted decisions.

1.4.6 Internalising costs to biodiversity

The task, then, is to demonstrate the value of biodiversity and its functions in ways relevant to business and governance. Then we can begin to internalise at least the most important parts of it into our economic considerations. And we can also persuade others, including our regulators and governments, of the need to do likewise for the common good of all in creating more stable future markets and a more truly 'rational' basis for trade in an increasingly environmentally responsible market. We'll look at this in the next section.

References

[1] Dasgupta, P., *The Control of Resources*, Harvard University Press: Boston, MA, 1982.
[2] Costanza, R., d'Arge, R., de Groot, R., Farber, S., Grasso, M., Hannon, B., Limburg, K., Haeem, S., O'Neill, R.V., Paruelo, J., Raskin, R.G., Sutton, P. & Van den Belt, M., The value of the world's ecosystem and natural capital. *Nature*, 387, pp. 253–260, 1997.

Part II   Biodiversity and business

2.1 People and nature

Our Victorian ancestors were entranced by the huge diversity of life. They curated collections of plants, seeds, shells, birds and their eggs, mammals and their hides, insects, diatoms (minute algae with ornate silica shells) and a variety of other organisms. They founded our municipal museums and zoos.

The irony, of course, is that this Victorian fascination with biodiversity also contributed to its substantive degradation. This was not merely directly through hunting and 'stamp collecting' of rarities and oddities for living and preserved collections. It was rather by inventing means of wealth generation that enabled the funding of these enterprises. This second indirect legacy is by far the most long-lasting, growing inexorably to threaten the very integrity of biodiversity at the global scale.

2.1.1 Our evolutionary past

Quite clearly, we are made of biological stuff and will ever remain a fully integrated element of the planet's biodiversity. Our cellular, sub-cellular and biochemical construction differs minimally, if at all, from those of the other animals and even plants with which we share this world. This is clear evidence of our common biological ancestry. As our technology advanced providing us with an ever-greater capacity to investigate our own nature, it came as a shock to many to learn that humans are, in fact, 98% genetically identical to chimpanzees! We are, in short, a small, recent and integral part of our home planet's biodiversity, special only in our more advanced capacity to comprehend, communicate about and manipulate the world around us, besides, perhaps, in our ability to conceive of ourselves as something apart.

Many of the attitudes and behaviours of modern culture imply that we see ourselves as something separate from the world that gave rise to us. The reality is that we are integral to the ecosystems of this planet that support the totality of our needs including health and well-being, wealth creation and spiritual and other spheres of fulfilment. These ecosystems contain the evolutionary wisdom of billions of years. Contrast this with the fact that modern civilisation first emerged around 5,000 years ago. This is little more than 200 human generations; 70 human lifespans of 70 years end to end. This should give us pause for some humility and a reappraisal of the depth of our own wisdom, and of our authority to act in ways that degrade biodiversity.

Yet, for all that, our special human attributes have allowed us to inflict changes on the natural world at a spectacularly rapid pace. Some of these changes may well be irreversible, given that they exceed the rate at which ecosystems adapt or evolve. While we cannot be sure – for we do not have the luxury of another 'control' planet

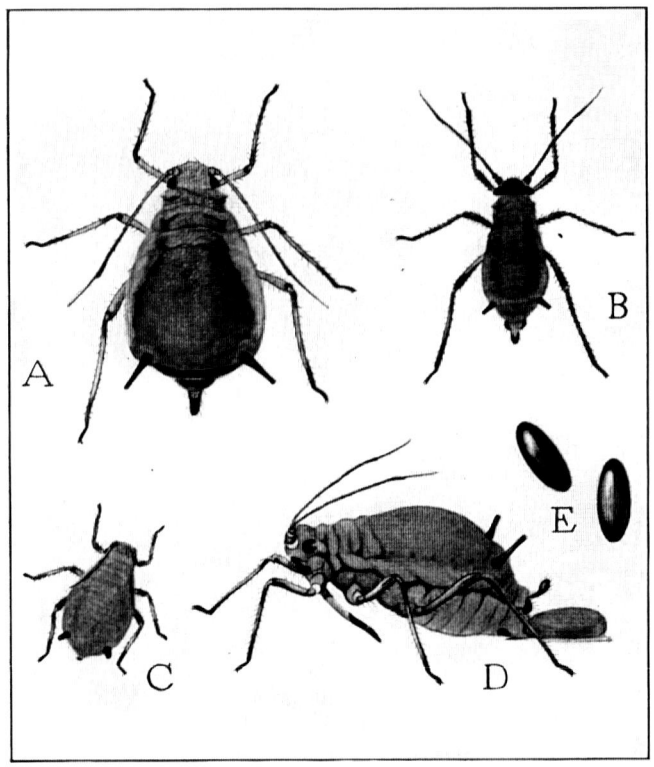

The green apple aphis (*Aphis pomi*)
A, adult sexual female; B, adult male; C, young female; D, female lay-
ing an egg; E, eggs, which turn from green to black after they are laid.
(Enlarged about 20 times)

to test out this assumption to breaking point – we seem bent on a course of cumulatively destroying biodiversity forged through billions of years, and degrading the very ecosystems on which we depend to sustain us indefinitely into the future.

For all our big brains and bipedal arrogance, we seem as yet not to have developed the foresight and common sense to prevent ourselves unravelling the very ecosystems that gave us life and can continue to support us in the future.

2.1.2   Population pressures

As far as successful species are concerned, we are probably no match for rats or cockroaches. But since we first appeared in Africa's Eastern Rift Valley, our capacities for innovation, communication and manipulation have allowed us to quickly colonise virtually every land mass on the planet. We haven't just survived – we've boomed, innovating our way around all the planet's natural mechanisms that might otherwise have controlled our population: predators, disease, food limitations and tendencies towards intra-species aggression.

It is to be expected that humans, like all other species, should influence the world of which they are part. However, the activities of our vast global population are placing huge pressures on the Earth's ecosystem and never more so than during our recent industrial development phase, during which we have come to consume more and more of the planet's biodiversity. We continue so to do at a scale that threatens its integrity and supportive capacities.

At the beginning of the Industrial Revolution (around 1760), there were only half a billion people on the planet. By the 1930s, global human population had boomed to two billion; in 2000, it exceeded six billion. Demographers expect a billion more people to be added to the world population between 2000 and 2030, almost all of this increase in cities in Africa, Asia and Latin America. Global population is projected to reach nine or ten billion on current trends before stabilising by around 2050. Today's population growth has no precedent, and is a largely unstoppable consequence of current driving forces.

Yet numbers alone do not tell the whole story. Each person has an individual potential of direct relevance to the achievement of a sustainable world. Disempowerment, loss of creativity or denial of opportunity wastes this human potential, and sows the seeds of social unrest and over-exploitation of resources. Poverty can therefore be measured not only in economic terms but also through access to resources. There are numerous indicators of equity of access to resources. For example, only 12.2% of the world's population can access the internet, while traffic jams in Brazil waste 200 billion litres of fuel a year (according to the Research Institute on Applied Economics). Today, 1.6 billion people have no access to electricity, and 2.4 billion rely on unprocessed biomass fuels – straw, wood, agricultural waste and dung – for cooking and heating. Worse still, it is projected that, in 30 years, 1.4 billion people will still have no electricity and 2.6 billion will rely on raw biomass fuels. Water shortages already debilitate much of the developing world, where water resources and their ecosystems are already over-stretched and

population is growing most rapidly. One worries about the impact of spiralling biomass combustion and water use on an already over-stressed planet.

2.1.3   Our industrial playthings

As our understanding of biodiversity grew, so too did our appreciation of its fragility. Ecosystems of the world have been changed by many things, including agriculture, deforestation, pollution, drainage, hunting, plant collection and introduction of alien species. European domination of the globe since the fifteenth century linked previously isolated parts of the world. This massively accelerated the pace of usage of biodiversity and other natural resources, driven harder still by the founding of the economic paradigm on which today's over-consumption is based. (The role of the economic systems in biodiversity loss is explored in later chapters.) To these already drastic impacts, we have to add the damage to indigenous biodiversity done by the many species of plants, animals and microbes that we have spread outside the areas in which they had evolved, a bad habit with which we persist both intentionally and inadvertently.

As long as we were limited by muscle power, most resource use was within the regenerative capacities of ecosystems. However, industrialisation automated many of the processes of extraction, use and conversion of biodiversity and other natural resources, displacing the patterns of low-intensity exploitation of preceding centuries. Humanity was now able to exploit the natural world at a much faster rate. Today, the pace of human-induced change is often many orders of magnitude greater (and accelerating all the time) than natural changes that take place in ecosystems over geological timescales. In short, biodiversity cannot cope with the pressures our species places on it.

2.1.4   Society's 'footprint'

Our burgeoning population places demands on the productive capacities of nature. There are serious question marks over the ability of our home planet to sustain this ever more populous and resource-hungry mass of humanity indefinitely. The WWF estimates that the Earth has about 11.4 billion hectares (about a quarter of its surface area) of productive land and sea space, after all unproductive areas of ice, desert and open ocean are discounted. Divided between a global population of six billion people, this equates to just 1.9 hectares per person.

WWF calculations suggest that the environmental footprint (resource demands normalised to an area of productive land) of the average rural African or Asian consumer was less than 1.4 hectares per person in 1999. This is in contrast with the average Western European's footprint of about 5.0 hectares and the average North American's footprint of around 9.6 hectares. Energy use accounted for the fastest-growing component of the global footprint between 1961 and 1999, increasing at an average rate of more than 2.3% per year. For the total human population of our planet in 1999, the average individual footprint was 2.3 hectares, 20% above the Earth's biological capacity of 1.9 hectares. In other words, humanity's demands now exceed the planet's capacity to sustain its consumption of renewable resources. The unbending laws of science tell us that we can maintain this global overdraft only on a temporary basis before inevitable catastrophe. And this is before we factor in runaway population growth, and marketing-driven aspirations of the 'developing world' to acquire the material commodities of the consumerist West. These basic analyses have been endorsed by those of other authoritative initiatives such as the Millennium Ecosystem Assessment, discussed in greater detail elsewhere.

If we cannot abate our consumption of the world's biodiversity, the whole global ecosystem on which we depend may simply crumble around us, unleashing untold human misery, perhaps even virtual wipe-out, in its wake. If every person on the planet today were to enjoy a contemporary European lifestyle, we would need the resources of three planet Earths. Clearly, if we are to avert catastrophe we have to either find two extra planets very quickly, or else innovate our way to vastly reduced consumption habits.

2.1.5   Putting a price on biodiversity

It has been a preoccupation of economists to put a monetary value on biodiversity and its attributes, usually expressed as value to society. (We have already looked at aspects of this in the preceding chapter.) There is some sense behind finding ways of internalising biological resources into the economy, such that their value is recognised in public policy and commercial decisions. However, as we have seen, it is far from certain that it is safe to substitute the non-market reality of nature with the man-made concept of financial value.

Economic valuation of biodiversity rests on a handful of techniques including replacement cost, substitution (e.g. loss of an aquifer substituted by the costs of treating and pumping water from other sources), 'willingness to pay' and contingent valuation wherein natural resources such as proximity of a river affects a property or other market price. While this last approach has some merit in recognising the value-added of less tangible factors on 'hard' market goods, most economic valuation mechanisms make dangerous assumptions. For example, since extinction snuffs out forever the legacy of billions of years of evolution, including their genetic wealth and functions with ecosystems, what is the market mechanism that allows us to calculate a 'replacement cost'? Or, if a water source is no longer usable and furthermore contaminates an ecosystem, is it always feasible to find another source of water and what is the knock-on impact on the ecosystems from which it is abstracted?

So, without going too much deeper into the complex and sometimes arcane world of environmental economics, let's just sound a precautionary note in applying monetisation of biodiversity as a panacea for all ills in the sustainable realignment of business. Nature has existed for billions of years, is infinitely complex, has created us and provides for the totality of our needs. So, to seek to sum up its 'value' in terms of the recent human invention of money is as arrogant as it is dangerous.

2.1.6   Cultural values

Aside from biophysical support and economic value, biodiversity is also important to humanity for far deeper reasons. From the gods of ancient Egypt to contemporary military mascots, various species have been seen as the embodiment of desirable qualities. This symbolism pervades many of the world's religions.

A prime example is the Shinto religion of Japan. ('Shinto' is a Chinese-derived word meaning 'the way of the gods': Shin = 'gods'; Tao = 'the way'.) In the Shinto belief system 'kami', or spirit, is found in all things, a different quality and presence of spirit being manifest in different entities be they living organisms, landscapes and even honourable human endeavours and the things that they produce such as buildings, swords or fine lacquerware. There is a variety of branches of Shinto, relating to different spiritual qualities ranging from the spirits of ancestors to any natural force or aspect of nature that inspires awe. However, the uniting factor of these branches and of the other diverse animist religions across the world is that natural forces have intelligence and an inexpressible value to humanity. This is clearly the case in pagan beliefs, and such attitudes still pervade modern Christian and other societies: witness the attraction of holy wells or springs and other such shrines.

The Hindu religion is as diverse as the races and landscapes of its adherents, as indeed is the Pantheon of major and minor gods that they worship. Yet this apparent pantheism masks the monotheistic principle at its core. Brahma, the creator, is the godhead. However, Brahma's work as the creator is done and the divine spirit of that act of creation manifests itself in all living things as 'Brahmin'. It is the devotion of Hindu believers to recognise Brahmin in nature and inspired human creations such as temples and shrines, and to do good works as an expression of that same spirituality, thereby building better *karma*. Around the Hindu shrines and temples of Deccan India, hunting or fishing is not permitted and the fish, fowl, monkeys and other living things are often fed, becoming almost tame. There are holy trees, holy groves, holy hills and places and holy rivers; the best known of India's holy rivers is the Ganges (or 'Ganga' in most Indian languages). Given the life-giving irrigation of the glacier-fed 'holy Ganga' across North India and Bangladesh, and the flush of water and nourishing silt borne by the Brahmaputra River, the 'son of Brahma', into its fertile basin across North India and Bangladesh, it is not hard to understand reverence for these 'eternal' rivers. In all, there are seven great holy river systems across the Indian subcontinent – the Ganga, Sarasvati, Sarayu, Yamuna, Narmada, Godavari and Cauvery – all of similar supportive value to humanity and all imbued with deep spiritual significance. Reverence for the spiritual content of nature and of the good acts that one can do is of the essence of the Hindu religion, whatever the faction.

The spiritual value of landscapes and ecosystems is commonplace across virtually all human civilisations. 'Earth Mother', sacred springs, wells or trees, gods both great and minor taking the form of animals, and a host of other imagery as ancient and broad as humanity itself speak of a depth of recognition and reverence for the world that supports and dwarfs us. Surely, we all feel this in our quieter moments alone in the wilds, whether we ascribe it a spiritual context or not? Surely, this is part of the renewal we feel when fishing, rambling, gazing out to sea, bird watching or otherwise finding excuses to 'hang out' in our natural habitat?

These clearly non-economic connections to the natural world are the cause of the outrage we feel when we see it destroyed or degraded by economic activities. This can have a powerful effect on the reputation of culpable companies.

2.1.7   Humanity and biodiversity

The biosphere of planet Earth is a system of almost infinite complexity and variability within which all things, both living and non-living, are intimately interconnected. As we have stated in various ways, all dimensions of humanity depend on it.

Every patch of earth or drop of water is host to a complex network of micro-habitats. Different organisms within ecosystems are adapted to differing regimes of temperature, nutrient concentration, light level and countless other variables, succeeding each other in activity, and often in dominance, throughout changing seasons, lunar phases and times of day. Yet, within this pattern of constant change, all parts of the global ecosystem interact to maintain the stability of the whole. Humanity, too, is an integral component of this whole-Earth system, affecting and affected by its changing state.

We need to ensure that the systematic loss of natural systems – aquifers, fresh-water systems, forests, soil, etc. – is abated and ideally reversed. We need to value the fact that nature is not just a fixed asset of 'nice green things' to look at and use. Instead, natural functions such as purification of air and water, formation of soils, provision of food and fibre and many more besides are the primary resources underwriting our long-term health, wealth creation and 'quality of life'. When we look at a natural resource such as land or river corridors, we have to recognise that we are seeing an irreplaceable and multi-purpose functional unit. Degrade its capacity to purify air and water, detain floods and retain water resources, produce food, support diverse ecosystems, maintain physical habitat, fertilise soils and provide economic and aesthetic potential, and we degrade our future in a surprisingly direct way.

The obvious corollary is that, if we protect it or restore it, we do the same for our own collective potential.

2.2   The business of business

We humans have a number of evolutionary advantages which, collectively, have helped us adapt to and populate all terrestrial habitats on Earth from polar to equatorial, wetland to desert, montane to coastal and island. We are blessed with the ingenuity to harness natural and human resources for our own benefit, but also to communicate and share the fruits of innovation and knowledge. Consequently, we can trace a rapid societal evolution from hunter-gatherer, through to settled farming, to industrialism and the information technology age. We are a learning species with unprecedented capacity to handle and innovate with complex information: we progressed from the Spinning Jenny to the Space Shuttle in just 200 years, and the pace of innovation only accelerates.

2.2.1 The origins of the Industrial Revolution

As society became differentiated, trading systems emerged along with the concept of money to enable the exchange of goods and services between the diverse pursuits within society. The farmer would trade food with the soldier, in return for defence. The builder would trade construction and maintenance of shelter with fishermen and fowlers, who traded back the fruits of wetland biodiversity. Philosophers and holy men traded ideas with weavers, who responded with clothing and baskets. And then, as civilisations diversified, we ended up with political, legal and financial institutions which also traded with the rest of society (though no one is still exactly sure what all of them offer us in return!)

In Britain, the Industrial Revolution, from the late eighteenth century (around 1760) through to the early nineteenth century (around 1830), brought unprecedented technological, socio-economic and cultural change. The country benefited from its relatively high degree of education, existing technological capabilities, merchant naval capacity and rich resources including coal, timber and metals. This created a change in social structure and habits that spread rapidly throughout what we now call the developed world.

Throughout this period, the manual labour that had up until then underpinned the economy, largely through working the land, was progressively replaced by industry and new machines. This change was fuelled, literally, by innovations in sources of power, predominantly from fossil sources such as coal, to power the new machinery built from the products of novel iron-making techniques. Ship, railway and canal building, the harnessing of water power and wider river engineering, the internal combustion engine and generation of electricity were just some of the advances that occurred throughout this compressed time period, each triggering or enabling further innovations. It was, literally, a revolution, as workers were freed from the land by automation to flock from the country into burgeoning cities and their factories, there to serve the wheels of industry. A new social order resulted, overthrowing long-established hereditary class-based and land-owning divisions within society.

Social enterprise changed from its focus on the land and the production of food and goods for basic survival; the new economy enabled grander visions of science, discovery and philanthropy. The new wealthy class undertook grand ventures such as investment in urban planning, public education, sanitation, sewerage, libraries and healthcare, laying the foundations for many social structures that remain with us today.

Step by step, individual inventors supported innovations in industrial processes such as spinning, textiles, metalworking and engineering, with progressively increasing dependence on machinery and less reliance on manual labour. As demand for human labour was displaced by the new machines, some saw a threat to employment, giving rise to a militant resistance movement of English workers whose name lives with us today: the Luddites. However, the momentum of industrialisation proved irresistible. Industrial innovation and the commercial gain that accompanied it have accelerated ever since, and we live today with a social system, resource use patterns, a market economy and corporate governance mechanisms that are the fruits of the Industrial Revolution.

At the outset of the Industrial Revolution, the supportive capacities and resources of the Earth were effectively perceived as limitless. Although local scarcities were encountered, such as mature timber from increasingly depleted British forests, new-found wealth from industry funded an age of empire-building when British interests overseas were often driven by access to regions rich in natural resources. Certainly, nature's capacity to absorb wastes was seen as boundless. Rather than resources being the constraining factor in the new economy, people were then perceived as the limiting factors in terms of both labour to service the new factories and customers for products. So, the baseline assumptions underpinning emerging economic and corporate governance systems were that nature's capacities were boundless but that humans were limiting. This precipitated an economic model that depended on stimulating consumption among a relatively low human population.

How times change! Both assumptions are reversed today. Human numbers are far from limiting. Instead, biodiversity and its services, critically including the resource of waste absorption, have been stretched beyond sustainable limits. However, in the face of this paradigm reversal, our economic and governance systems remain predominantly rooted in wasteful resource use habits seeded in the early days of industrialisation and shaped by these anachronistic assumptions. Degradation of biodiversity through over-exploitation and pollution, reinforced by the inequitable distribution of resources across society, is an inevitable impact.

Consequently, 'progress' as driven by inherited economic development pressures within today's industrialised and industrialising nations often breaks the nourishing links between biodiversity and business. Biodiversity loss is seen, or more commonly unseen, as the price of economic development, with a blind eye cast in the present to the inevitable consequence of impoverished biodiversity inhibiting human and economic potential in the future. Almost inevitably, the greatest burdens of this loss fall on poor and marginalised people, who rarely have any voice in the governance of business and government strategies nor a fair share of their proceeds.

2.2.2   Economics as the sole driver

The global economy expanded sevenfold in the last half of the twentieth century, with output rising from $6 trillion of goods and services in 1950 to $43 trillion in 2000. Growth in the world economy during 2000 exceeded that achieved during the entire nineteenth century. Yet what does the transfer of all this money actually mean, and how does it map back to the biodiversity and human well-being that underpin prosperity?

If we view nature as limitless, then we can afford to pour concrete over it forever, and to over-harvest from marine fisheries, aquifers, forests and other resources. We can afford also to transfer alien species from one geographical zone to another, taking them out of the ecosystems with which they have evolved as interdependent functional components. Once let out of the bottle, the genie – or perhaps we mean the genes – can run amok in these new environments. We can also afford to be cavalier in our use of agricultural land, regardless of the consequences of soil erosion, contamination by agricultural inputs, displacement of species and ecosystem services and a host of other hidden costs.

Yet, as we observed in Part I of this book, problems inevitably arise when we break the link between the economy, the biodiversity it depends on and the society it is intended to serve. Then, businesses act rather like rampant cancer cells, serving their own growth and generating wealth in ways that destroy the host biological organism; they cease to respect their role in servicing the needs and well-being of the human society that formed them. When the rampant economy enforces this behaviour, the servant becomes a tyrant that all must serve until their capacity to do so is stretched beyond breaking point. Erosion of biodiversity, life-support mechanisms and human potential are inevitable consequences.

As discussed already, the reality is that money is a human-made device for trading, and has no absolute value of its own, whereas some economists argue that everything can be monetised. In the 1990s, there was an often-used saying that 'The business of business is business', traceable back to the US economist Milton Friedman, which argues that peripheral concerns such as environmental or ethical issues are a dereliction of the duty to maximise value for shareholders. In other words, commercial enterprises should preoccupy themselves solely with wealth creation by any means, leaving any 'burden' of red tape to regulatory bodies and ignoring the whingeing 'greens'. Few among the hardest-nosed of business leaders today would define their role in society in quite so extreme a fashion. The world is changing. However, the prime regulatory requirement imposed on directors of businesses is to maximise value to shareholders; this is hardly the best leadership that government could offer, particularly when it conspicuously fails to back up its calls for businesses to report on environmental performance with statutory compulsion. However, business leaders are generally far from stupid or short-sighted, and most well understand that we live in a world of changing values, looming threats and shifting market conditions. Sustainable profit will depend on a different relationship with biodiversity and people than that of the 'greed is good' 1980s or the environmentally illiterate phase of industrialisation that preceded it. There is a widespread recognition in business of the need for innovative approaches to making a return on investment, including accounting for interactions with biodiversity.

Business, then, has to find a new master if it is not to act with the destructive force of a poorly evolved parasite on the natural world and society. It has to recognise demands and impacts on biodiversity, reclaim its role in profitably serving society's needs, and reform itself into an agent of sustainable forms of development. This is not altruism; it is enlightened self-interest in a more aware, responsible and fast-changing world.

Business is changing because the world that gave rise to it has changed radically. Tomorrow's business will have no option but to take account of it, and today's leaders will want to pre-empt that inevitable change.

2.3   Biodiversity and sustainable development

Our industrial and agricultural innovations have provided us with unprecedented material quality of life, expectation of health and longevity and population levels. However, they have also imposed a burden on biodiversity through over-exploitation of natural resources, pollution, physical disturbance of the landmass and ocean floor, and reduced capacities for self-purification of air and water. In so doing, they have reduced the capacity of the natural world to support our needs and aspirations in the longer term.

The implications of human over-exploitation of biodiversity are serious for the integrity of global ecosystems, their efficient functioning and the capacity of people to achieve their potential.

2.3.1   Nature in mind

Rachel Carson's seminal book *Silent Spring* [1], published in 1962, was an almighty shock to a world that was largely blind to the potential of its actions to harm biodiversity. *Silent Spring* documented the detrimental effects of chemical pesticides; it alerted us to the fact that the damage can be both long lasting and also often remote from the point of chemical use. The book integrated evidence of pesticide accumulation in polar birds, the contribution of body loads of pesticide residues to mass deaths of farmland birds and a range of other environmental and human health issues. It was a stark warning.

The warning was reinforced by a sequence of high-profile environmental disasters during the 1960s and 1970s. The *Torrey Canyon* tanker spill (England), Love Canal (the US), Minamata Bay (Japan) and many others besides are part of a shameful catalogue from these decades. All of this was, of course, con-textualised by the first images of our small, vulnerable-looking home planet in the darkness of space as beamed back from the Apollo 8 space mission in December 1968.

Awareness of the vulnerability of ecosystems to human misuse and the dawning of awareness of the need to live within the landscape in ways that protect ecological integrity have a pedigree as far back as Aldo Leopold's call for a 'Land Ethic' in his 1949 book *A Sand County Almanac* [2]. Some also see the dawning of consciousness about human degradation of rivers in Harry Plunket Greene's evocative *Where the Bright Waters Meet* in 1924 [3]. However, broad public awareness and concern about the loss of biodiversity only began to frame significant government-level responses from the 1970s.

Modern thinking about more sustainable relationships with living landscapes were first enshrined in an international agreement under the Ramsar Convention (the Convention on Wetlands of International Importance Especially as Water-fowl Habitat) in 1971, which proposed the 'Wise Use' of aquatic and wetland ecosystems. The Ramsar Convention, signed in the Iranian city after which it was named, recognised that historic approaches to conservation by fencing people out of nature reserves – so-called 'fortress conservation' – were simply infeasible in the long term. There are moral and practical reasons for this, and it is particularly the case for wetlands. The high productivity and generally flat topography of many of the world's wetlands render them valuable to society in supporting diverse ways of life and economic activities. 'Wise Use' under the Ramsar Convention, now agreed to be synonymous with the concept of sustain-able development, requires that human activities be adapted (as many traditional land practices already are) to exploitation of wetlands for human gain in ways that protect their natural character, including biodiversity and the functioning of the ecosystem.

The international Stockholm Conference on the Environment in 1972 – the first in what has turned out to be a series of decadal UN-sponsored international gatherings to define and respond to the growing global environmental challenge – gave a kick-start to the fledgling 'environmental movement'. The next significant milestone was the UN's World Conservation Strategy (1981) [4], which made explicit the indivisible linkage between biodiversity and human exploitation for social and economic progress.

The UN then set up the World Commission on Environment and Develop-ment (WCED), more famously known as the Brundtland Commission after the Norwegian Environment Minister (and later Prime Minister) who chaired the international panel convened to address the growing threats to the future of human development. The WCED's report of 1987, titled *Our Common Future*, communicated in clear terms for a global audience the indivisible connec-tion between environment, economics and society, and how this related to the possibilities of evolving beyond the dominant industrial paradigms [5]. *Our Common Future* introduced the concept and term 'sustainable development' to a wider world.

Successive major UN conferences on sustainable development followed to monitor progress and plot future strategy. At the Rio de Janeiro 'Earth Summit' in 1992, the Convention on Biological Diversity was one of a raft of global agreements signed. International intent to halt the accelerating loss of biodiversity by 2010 is articulated in a commitment under the Convention on Biodiversity. The Johannesburg World Summit on Sustainable Development in 2002 endorsed these and set further targets, including some relating to reversing the continuing decline in biodiversity and offering greater assistance to the developing world to meet its needs. The 2010 target was also endorsed by the EU in 2002 as the Gothenburg Target, which has implications for all aspects of societal activity and from which further economic instruments and regulations can be expected to flow.

Have we done enough? The current state of the Earth's biodiversity reflects the consequences of our actions and, at the same time, the capacity of the planet to continue to allow us to meet our needs. It suggests that the work has barely begun.

2.3.2   The evidence of biodiversity loss

The evidence is unambiguous. We have only just embarked on the journey of slowing the rate of damage to the natural world, let alone the longer path to benign and sustainable human development or the recovery of ecosystems.

The UN's Millennium Ecosystem Assessment (MA), described already, was surprisingly (or perhaps not) met with a general lack of interest among the world's media when the synthesis of its general conclusions was published in 2005 [6, 7], despite it being the product of 5 years of research by more than 1,300 scientists from 95 countries, with conclusions emphatically and authoritatively underlining that human activities have changed the world more rapidly over the last 50 years than during any other time in history. Yet, for all the advantages this has brought for some, billions of people remain in poverty. The MA put numbers on the 'overshoot' of the demands that industrialised society places on the finite supportive capacities of global ecosystems, and defined the breaking points that these self-same ecosystems have reached at our hand.

The MA highlights the many ways in which humanity's destructive patterns of development are severely compromising the integrity of the natural world and its capacity to support us into the future. Our collective contribution to the liquidation of the natural world threatens to derail the United Nations' Millennium Development Goals to halve global poverty and hunger by 2015.

Political leaders have been as coolly unreceptive as the media despite the implications of the stark, scientifically grounded MA. More substance is added to the MA's already weighty findings by the fact that they endorse the already dire warnings of the WWF's excellent series of *Living Planet Index* reports [8], which have catalogued the decline of global ecosystems and the spiralling demands of modern society on them. Add to this the findings of the long-running series of UNDP *Human Development Reports*, which analyse the impacts of inequitable sharing of wealth and power and the implications of systematic environmental degradation for peoples both poor and rich, and we could have no clearer warning of the consequences of our current pathway of exploitative industrial development [9].

2.3.3   Where now for nature and humanity?

So society has become aware, albeit sometimes grudgingly, of the threats to its future. It has also initiated some international agreements to action. However, the inertia of vested interests and ingrained habits, corporate and governmental, are as yet impeding a response proportionate to the threat, currently constraining it to actions that are too marginal, fragmented and tardy. Lofty commitments are necessary, but have to be backed up by appropriate legislative signals including eventual sanctions and/or incentives.

Declining human security and potential are the only logical outcomes if we cannot muster the foresight and courage to change from our recent pathway of industrial development.

2.3.4   A brief history of conservation

Our first attempts at conservation of biodiversity were based on captive populations and the designation of reserves that acted as 'living zoos'. While the need for nature reserves with restricted access is still essential to protect the most critically endangered of habitats and species today, the wider emphasis has since evolved from the 'fortress conservation' model. This is in step with the growing realisation that nature and people are not separate. Keeping humans and nature apart simply cannot work in an overcrowded and ever more populated world, and is anyhow an unethical position if it stops people from meeting their needs.

In order to protect certain vulnerable species, rather than looking exclusively at local habitats, there has been a need to control international trade in them, their parts and derivatives. International agreement to CITES, the Convention on International Trade in Endangered Species, was a milestone in controlling the impacts of international trade on national or regional conservation. (CITES is examined in greater detail in Part IV.)

Agri-environmental subsidies were also introduced to bring land use into greater sympathy with the needs of biodiversity. Their emphasis was initially on payment for 'profits foregone', where land managers were paid for agreeing to discontinue practices harmful to nature. However, as thinking about resolving the conflict between biodiversity and agri-business developed, this approach became perceived by some influential commentators as a form of blackmail: society agreed to pay farmers for not doing things that harmed the environment and its capacity to support wider human needs!

Through the 1990s, agri-environmental payments were gradually realigned to reward management practices of benefit to the conservation of nature and some aspects of heritage while simultaneously supporting profitable farming. Since 2005, with the introduction of the Environmental Stewardship agri-environment regime in the UK, consistent with recent changes in the EU's Common Agricultural Policy (CAP), subsidies are explicitly tied to benefits for society from land management. Notwithstanding a legacy of payments based on land area, all former linkages of agricultural subsidy to commercial outputs have been scrapped.

The regime is changing slowly to recognise that land, water and other habitats exist for purposes other than the commercial production of food and other commodities for the sole benefit of local landowners. However, the very division of subsidised from non-subsidised land use perpetuates some aspects of the old 'fortress conservation' model in that concern for nature's needs remain hemmed into those locally defined areas that attract such subsidies. Meanwhile, the generality of land use is implicitly assumed to continue to erode nature and ecosystem functioning.

2.3.5   The origins of the concept of sustainable development

By merely prohibiting access to areas of habitat, we may create reserves for some species while, outside the barbed wire, their destruction carries on unabated. Furthermore, there are human rights issues associated with the displacement of people whose livelihoods depend on exploitation of those habitats, and whose traditional wisdom might just embody practices sympathetic to co-existence with biodiversity. The mass displacement of people during the establishment of the Kruger National Park and other nature conservation areas in South Africa is not a model of which anyone is proud today, least of all the marginalised generations who have since grown up in squatter settlements. In South Africa and other nations now seeking to redress the situation for 'historically disadvantaged individuals', there is in fact a growing need to develop the kinds of 'unproductive' land often retained for nature reserves as well as other types of land. The challenge is, therefore, to find ways of using land that can raise these people from poverty while maintaining biodiversity and its many beneficial functions, particularly as they relate to river catchment systems.

Without innovation, rapid population growth and associated legitimate demands for resources will inevitably swamp attempts at traditional conservation. There has therefore been a progressive recognition of the pressing need to replace the flawed notion that we can fund 'conservation' with a small proportion of the profits arising from the systematic degradation of nature.

Effective conservation lies instead in finding means by which human activities can interact with the myriad other life forms on which they depend without destroying ecosystem integrity or function. It will require novel forms of wealth creation that support societal needs without unravelling the web of life that sustains society. As described above, the Ramsar Convention began to change the way international society looked at the challenge of nature conservation, and the Ramsar principles became widely acknowledged in the growing international quest for a mutually supportive alignment between biodiversity, society and business.

'Sustainable development', the term brought into broad public consciousness by publication of *Our Common Future* in 1987, is about a pathway of development that addresses the simultaneous protection or enhancement of environment, economy and society. A great deal of thinking and much practical work has been undertaken since to try to bring this shared global vision to reality, both locally and at wider scales.

We have little if any time left to make a broader, society-wide transition. The negative trends in biodiversity discussed above suggest that we have a way to go yet before we can claim any serious progress.

2.3.6   Developing sustainably

Many current issues – declining biodiversity, climate change, fishery collapse, erosion, deforestation, inequities between and within nations, water shortages and many more besides – exemplify the way in which stark warnings of environmental collapse and ensuing human misery are still ignored by the very political systems intended to advance the common good for the people they serve. For all the international agreements and political rhetoric, the reality of the developed world today is that short-sighted consumption habits remain not only in full flow but in growth, reinforced by vested interests unresponsive to the need for action in the face of overwhelming evidence.

We now stand at a watershed in history. Behind us is a legacy of innovation that, while emancipating many of us in the present, has been largely blind to its own adverse consequences for the natural and human resources underpinning long-term well-being. Ahead of us lies an opportunity to live more sustainably, by unleashing the formidable power of human innovation to plot pathways of continuing development that account for what we now know, and will continue to learn, about the natural resources on which our personal and collective futures depend. Now, at least, we still have the possibility of making a positive choice.

In reality, we have stood at this watershed for some decades, watching the window of opportunity narrow by the instant. Faced with the horrors implied by continuing unsustainability, we have no choice but to elect for a pathway of sustainable development. Appeals from vested interests for a moratorium on substantive action due to lack of proof of any problem, and of the efficacy of likely solutions, is no longer itself sustainable. For our own well-being and economic success, as well as that of the long-term future, we can no longer afford the luxury of prevarication.

Sustainable development is integral to the next phase of human evolution, critically including finding alternative ways of meeting our needs and making money without all of those adverse consequences for biodiversity.

References

[1] Carson, R., *Silent Spring*, Hamish Hamilton: London, 1962.

[2] Leopold, A., *A Sand County Almanac: And Essays on Conservation from Round River*, Oxford University Press: New York, 1949.

[3] Plunket Greene, H., *Where the Bright Waters Meet*, Medlar Press: Ellesmere, UK, 1924 (2007 edition).

[4] IUCN/UNEP/WWF, *World Conservation Strategy: Living Resource Conservation for Sustainable Development*, Gland, Switzerland: IUCN, 1981.

[5] WCED (World Commission on Environment and Development). *Our Common Future*, Oxford University Press: Oxford, UK, 1987.

[6] Millennium Ecosystem Assessment. *Millennium Ecosystem Assessment*, 2004, www.maweb.org.

[7] Millennium Ecosystem Assessment. *Ecosystems and Human Well-Being*, Island Press, 2005.

[8] WWF, *Living Planet Report 2004*, WWF: Godalming, UK, 2004; http://www.panda.org/downloads/general/lpr2004.pdf).

[9] UNDP (United Nations Development Programme), *Human Development Report 2004: Cultural Liberty in Today's Diverse World*, UNDP: New York, 2004; http://hdr.undp.org/reports/global/2004/).

2.4   The changing business agenda

Sustainability is the new business agenda, defining the markets that business must serve in supporting human needs and demands in ways that cease to erode biodiversity. This bold statement is becoming a fact, and will increasingly do so as our inherited resource use habits and mushrooming population cut ever deeper into nature's supportive capacities. Without serious realignment, they will limit society's freedom to operate and ability to achieve its potential. The advantages of addressing any kind of market change proactively, rather than being forced into reactive response further on down the road, will be clear to anyone in business.

2.4.1   Looking backwards and forwards

This may well be a bold assertion, but think back 20 years. Who cared or had even heard of many of today's pressures, far less thought that they might ever impinge on business? From climate change to usage of water and energy, consumption of timber and paper, reputation issues from the environmental and ethical impacts of sourcing of food, gravel and imported products and recycling targets and rising waste costs. Many of the chemicals in common use 20 years ago have since been banned, phased out or subject to stringent controls due to impacts on biodiversity and human health.

Environmental reporting is now commonplace, and energy and water ratings are applied to domestic appliances. Emission limits imposed on factories and cars have never been more stringent and are becoming tighter all the time, while unwelcome media attention can be assured for a business with a negligent or poor environmental record.

2.4.2   A changing world

Rock-solid evidence of how the biodiversity concerns of a changing world bite harder on business every day is seen in the catalogue of changing regulations with which companies have to comply. A scan of just a few of these that have emerged only since the turn of the millennium should both alert us to the threats of ignoring our responsibilities, as well as hint at the new opportunities that a strategic response may create. This summary has a UK and European focus, but we see a similar pattern across much of the world.

Pressure on water resources has been evidenced in the early years of this millennium by Drought Orders in some localities in the UK, as well as escalating water service charges, compulsory domestic metering and the recommencement of dialogue about a new reservoir in the Thames catchment. Meanwhile, there is a continuing severe drought in Australia, monsoon rains are becoming less predictable and less abundant in India, and South Africa is facing increasing water scarcity. Some of these water issues are caused by a changing climate, but most are exacerbated by degradation of the water storage functions of the local catchments.

Controls on spatial planning and infrastructure are governed by increasingly stringent Policy Planning Statements (PPSs), the Countryside and Rights of Way Act 2000 and the implementation post-millennium of the Town and Country Planning (Environmental Impact Assessment) (England and Wales) Regulations 1999 in the UK. All of these increasing strictures on forms of development are due to growing concern about potential ecological and environmental impacts. At European scale, the EU Strategic Environmental Assessment (SEA) Directive, adopted in 2001 and transposed into UK law in 2004 and 2005, requires a formal environmental assessment of certain plans and programmes that are likely to have significant effects on the natural environment. All such instruments highlight the growing attention in the UK and Europe on biodiversity, ecosystem functioning and other environmental considerations in development.

The international commitment to halt the decline in biodiversity by 2010 under the Convention on Biological Diversity, endorsed at the World Summit on Sustainable Development in Johannesburg and also agreed as the Gothenburg Target by the EU in 2002, makes a binding commitment to significantly reduce or halt the rate of biodiversity loss. This has implications for all aspects of societal activity, nationally, continentally and through trade and other relationships across the global community. We can confidently expect new policy measures, from more stringent regulations to incentives for behaviour change with respect to biodiversity, to be brought to bear on business in the years ahead.

The Aggregates Levy was introduced in the UK in April 2002, imposing a 'quarry gate' tax on mined aggregates. This levy has since increased progressively. It too stems from concern about the various impacts of quarrying activities on natural and human environments, with the Aggregates Levy distributed through the Aggregates Levy Sustainability Fund (ALSF) into local environmental and community schemes to offset any such harm.

In part related to addressing pressures on biodiversity, we have seen the introduction and incremental tightening of a new generation of post-millennial European environmental regulation containing 'take-back' and 'producer responsibility' as operating principles. These include the EU Waste from Electrical and Electronic Equipment (WEEE) Directive (2002/96/EC); the EU End-of-Life Vehicles Directive (2000/53/EC); and the EU Packaging and Packaging Waste Regulations (94/62/EC). These instruments also contribute to reducing waste streams, an intended goal of additional EU Directives on Waste to Landfill (implementing regulations came into force in England and Wales in 2002, 2004 and 2005) and Waste Incineration (applied to new plants since 2002 and to existing plants since December 2005). In addition to European drivers, the UK Landfill Tax currently imposes a charge, currently £21 per tonne on non-hazardous waste but set to rise above inflation rates, as an incentive for re-use or recovery. This is stimulating growing markets in recycled materials. These collectively contribute to reducing releases of mined and synthetic chemicals, reducing 'take' of virgin resources from ecosystems and a more efficient and equitable sharing of resources across society. Biodiversity impacts from resource extraction and waste disposal, including 'land take' for mining and other processes as well as landfill, are part of the driving forces for the more efficient and sustainable resource use patterns stimulated by this raft of European and domestic legislation.

It would be foolish to expect no additional pressures of this kind in the coming years. As the Millennium Ecosystem Assessment and other metrics of our impacts on biodiversity warn us, things are substantially worse than 20 years ago, and are continuing to worsen. Biodiversity concerns and, more broadly, sustainable development look set to become *the* business driver of the future. The above selection of opportunities and restrictions emerging over the first few years of this millennium – and this only a small sample of more widespread change – provides compelling evidence that, even over this short timescale, sustainable innovation offers a return on investment and a hardening business case for addressing biodiversity concerns. Equally, those in denial or acting as market followers to sustainability pressures are incurring the increasing costs of reputation loss and *post hoc* reaction to the shifting markets and regulations of a changing world. The market will inevitably continue to change, with biodiversity and wider sustainability pressures among the primary driving forces. The evidence of cultural change, cemented in legislation, is as plain as the implicit benefits of responsible business.

The benefits of proactively meeting the needs and markets of the inevitable changes in our world should be self-evident. Determining and modifying the relationship between business and biodiversity is an overwhelming necessity with clear economic implications, in terms of both the negative consequences of inertia and the positive benefits of action to pre-empt future markets. The bottom line is that all these societal concerns are becoming internalised into both the market and the regulatory framework at a rapid pace. You're going to have to respond to them anyhow, so why not make it a strategic and profitable commitment?

2.4.3   Biodiversity in the context of sustainable business

The complete dependence of businesses on biological diversity does not appear to be fully appreciated. Modern industrial agriculture, for example, has been incentivised throughout much of the latter half of the last century, often against the preferences of farmers themselves, to treat the fertility and characteristics of the soil either as a given or as an irrelevance, using man-made fertilisers and pesticides to maximise productivity regardless of the suitability of the landscape and the consequences that ensue. Despite some laudable initiatives, by and large the musical instrument industry still relies heavily on endangered tropical tonewoods and timbers. In the back pages of the Sunday papers, it is commonplace to find glossy adverts for luxury home and garden furniture plainly made from timber from threatened tropical forests. The general assumption in manufacturing industries still appears to be that resources originate from suppliers, the tap or the National Grid and not from the natural systems from which virtually all are ultimately derived.

The bulk of business has much more progress to make in waking up to the reality that biodiversity underpins economic activities. As we have seen, wealth generation is, ultimately, merely the manipulation of ecosystem 'goods' and 'services' through human labour and creativity. If business can accept that it must invest its financial capital wisely if it is to be sustained, should this principle not also apply equally to the ecological capital at the root of profitability?

A number of leading businesses are waking up to the fact that unifying driving forces underlie the apparently diverse pressures of a fast-changing world, and that these do indeed pertain to a natural world rapidly dwindling in resources due to the pressures of a resource-hungry and growing human population. Unavoidably and increasingly, these sustainable development pressures will become the defining features of our future and the marketplace. A robust understanding of sustainable development, and our relationship with the biological diversity of the Earth, therefore becomes a matter of increasing our capacity to predict, pre-empt and position ourselves best to profit from the inevitable changes. In effect, this is nothing more than prudent and informed risk management.

Biodiversity defines the capacity of natural systems to support business. Other factors too define the 'headroom' for business to succeed. For example, economic returns assured by the meeting of the needs of the society, ethical trading and responsible production all contribute to the general 'licence to operate' granted to business by society. Although biodiversity is one of the most fundamental concerns of business going forwards, it is unwise to separate it out from consideration of social impacts and economic success. Ecosystems, social systems and economic well-being are in reality an indivisible triad that must be addressed simultaneously.

This realisation presents a terrific opportunity to business. A true commitment to sustainable development is an investment in tomorrow's economy, a spur to innovation and a compelling reason for the protection (or restoration) of biodiversity and its supportive capacities. However, it can be confusing to try to think through all the good things that we know make sense: issues such as recycling, ethically traded coffee, environmental management systems, trading relationships protective of biodiversity and a host of other single-issue ethical and social matters. To achieve an integrated approach, we need a single cohesive framework through which to apply this holistic worldview into the many practical strategic and operational decisions in business, and to do so in a manner consistent with business processes. (We'll turn to this in Part V of this book.)

2.4.4   Preparing for a changing world

The natural world that we inhabit, and on which we depend absolutely for our physical, spiritual and economic well-being, is in rapid transition due to the consequences of our collective actions and particularly our economic and industrial systems. It has been so for millennia, though the pace of change has accelerated since the onset of industrialisation towards the end of the eighteenth century in the Western world. It is now intensifying at unprecedented rates.

Although industrialisation created unprecedented levels of wealth and public health for the privileged, much of the unintended human-wrought change has

been far from good news for nature and society at large. Marginalised and poor communities often suffer disproportionately from degradation of ecosystems and an associated reduction in life opportunities, and to these people we can add future generations. We have been slowly realising the consequences of these anthropogenic changes over several decades, and awareness is dawning about the likely prognosis if we do not change our ways.

However, business is waking up to the need to address its interdependence with biodiversity, and the benefits that can accrue from taking responsibility for it. In Part III, we'll look in more detail at some of the implications for business and society where biodiversity is exploited to breaking point. Then, in Part IV, we'll explore what can practically be done before turning, and in Part V, how to integrate these measures into a unified corporate strategy.

Part III   Beyond the limits

3.1   Biodiversity pushed too far

It is at this point that the biodiversity and sustainability advocate usually dives into the doom and gloom of how we humans have harmed the natural world, and thereby prejudiced our continued well-being and livelihoods. And on that score, I am not going to disappoint throughout this part of the book!

Given our pathway to the present, it is inevitable that any serious read-out of the state of our world is prone to be a little depressing. However, I am offering a very brief overview of ecosystem damage purely in order to set a context for what is to follow, and as a foil for the inspiring and self-beneficial measures that can flow from responsible and far-sighted management of the relationship between business and biodiversity.

Where human exploitation is within the 'carrying capacity' of supportive ecosystems, respecting natural limits beyond which degradation of their character, integrity and regenerative capability will ensue, a sustainable balance has been struck. This can, by definition, continue indefinitely. However, where human activities exceed such sustainable limits, the integrity and long-term viability of the ecosystems and resources in question are automatically placed in jeopardy along with their capacity to provide for human needs.

In the following few chapters, we'll explore different examples of where business and society has placed too great a burden on biodiversity, with inevitable adverse economic, social and ecological consequences.

Don't worry: I will move on to the positive examples and opportunities in Part IV.

3.2 All at sea

The European explorer John Cabot made the first recorded landfall in Newfoundland, Canada, in 1497 while seeking a new trade route to the Orient. However, Cabot's accidental discovery proved far more lucrative in the long term, his crew reporting that they merely had to lower a basket into the sea and it would come up full of cod.

3.2.1 The Grand Banks

This marked the beginning of a 500-year fishing industry on the Grand Banks, which came to dominate Newfoundland's economy. In the centuries that followed, abundant fish stocks drew many people there to harvest them with small boats, the limitations of their manual efforts constraining exploitation within a sustainable limit.

Today, by dramatic contrast, there are no fin-fish fishermen to tell these tales, and precious few cod either. The virtual overnight collapse of the Canadian cod fishery through over-fishing, with the loss of the tens of thousands of jobs that depended on it, is a dramatic tale of lack of foresight, continuing greed in the face of clear evidence and poor decision-making. Despite a ban on all cod fishing on the Grand Banks in 1992, cod stocks are still to recover.

3.2.2 The industrialisation of the fishing business

The preceding centuries of low-intensity fishing made little impact on the productive ecosystem of the Grand Banks fishery, a source of sustainable livelihood for thousands of people. Yet, from the 1950s, increasingly massive factory trawlers, despatched predominantly from Russia, Europe, the United States and Canada, began to net the fish and denude the sea bed at increasingly unsustainable rates, systematically emptying the Canadian Grand Banks of cod.

The situation worsened through the 1950s and 1960s as trawler size and fishing power quickly grew, and the newly designed trawlers, modelled on factory whaling ships, permitting round-the-clock harvesting. These massive factory fishing ships could haul in as much as 200 tons of fish every hour, all of it quickly processed and deep-frozen. The cod population was dented, but fishing efforts increased until the total cod catch peaked at 800,000 tonnes in 1968.

When the recruitment of new generations into the cod population and impacts on the marine habitat that the fish depended on were assessed, the rate of take from the fishery was quite clearly demonstrated to be unsustainable. However,

the scientific facts and science-based warnings were not heeded to any serious extent. By 1975, annual catch had dropped by more than 60%, due to dwindling stocks rather than any attempt to manage resources, and populations of other species were also found to be plummeting.

Canada extended its fishing limit from 12 to 200 miles in 1977, a potentially progressive conservation measure, seeking to keep out foreign competition and reserve the remaining fish for its own fleet. By this time, however, the fishery was already in steep decline.

3.2.3   DIY madness

If Canadian policy had stopped there, the fishery might have recovered. However, the government instead ploughed huge investment into its own fishing industry, effectively replacing foreign mega-trawlers with home-grown boats which, unsurprisingly, continued to destroy the cod stocks, the seabed and the whole delicate marine ecosystem. To compensate for declining catches, the factory trawlers adopted sonar and satellite navigation to target the fewer remaining large cod shoals, which were particularly easy to exploit when grouped in the breeding season. Cod catches remained steady throughout the 1980s as increasingly powerful and more sophisticated technology chased fewer fish, rendering policy-makers deaf to the scientific warnings. After all, the value of the fishery in terms of jobs and revenues was too great to disregard, and it seemed that the well could never run dry.

The consequence of all this short-sightedness was, sadly, all too predictable. By 1992, the cod population had hit an all-time low, at an estimated 1% of the levels of the 1960s. This forced the government to close the fishery, putting 30,000–40,000 people out of work, robbing many fishing communities of their livelihoods and seeding a culture of poverty, depopulation and domestic breakdown.

A decade and a half on, stocks are yet to recover. Indeed, there is some pessimism about whether they ever will, thanks to the systemic and sustained harm inflicted on the whole Grand Banks ecosystem. As far as Newfoundland is concerned, the world's most prolific fishery is history, wiped out forever through greed and stupidity.

3.2.4   Our own backyard

It is easy for us to look back, both across the cool Atlantic and over time, to point to the errors of others. Yet we in Europe today seem to have learned little from the Grand Banks experience. Our scientists are issuing the same types of warnings, which are challenged in the same way by the politically powerful lobbies of European commercial fishing interests on the grounds that more investment of energy and technology is still yielding them cod.

The European and Canadian Grand Banks fisheries show marked similarities. In both, the groundfish, particularly cod, are (or were) the foundation of the fishery.

Today, in Europe, we are seeing many parallels with the Grand Banks fishery before its precipitous decline, including the same 'hyper-aggregation' behaviour among the fish as they cluster into greater densities when the environment comes under increasing pressure. The behaviour of the fisheries and the politicians mirror each other across 'the Pond' as our leaders seem wholly preoccupied with hag-gling with their foreign counterparts at periodic European fishery policy meetings for a greater share of a dwindling – in all probability disappearing – resource.

We can only hope that common sense and the counsel of robust science will prevail before our domestic fisheries crash from mere depletion into complete non-viability. Disastrous though loss of these stocks may be, we have also to be conscious that cod populations are merely the tip of the proverbial iceberg of wider degradation of the entire oceanic ecosystem. With this comes the loss of a whole network of ecosystem services – the essential provisioning, regula-tory, cultural and supporting services as defined by the Millennium Ecosystem Assessment – the benefits of which support the needs, aspirations and security of all of global society.

3.2.5 All at sea

The situation is, sadly, replicated across the whole of the world's oceans. It is commonly agreed that somewhere between two-thirds and three-quarters of oceanic fisheries are now being fished at their sustainable yield or beyond, or else are recovering from depletion. Many are collapsing, or have already done so. And this when the science base necessary for the calculation of sustainable yields has been available for decades to guide our political leaders and policy-makers. The 1997 collapse of the highly valuable hake fishery off Argentina in the South Atlantic is just one of a catalogue of lessons still largely to be heeded.

Many fisheries, and certainly those across Europe, have caps of maximum allowable catch, allocated by nation and subsequently broken down to active fishing boats. However, even here, the quotas that are set are highly influenced by politics, to the point where the underpinning science is a bit-part player in the all-important game of gaining economic advantage over competitors. We have, it seems, a compulsion to take more and more shares of a resource that our collec-tive greed is destroying. 'The tragedy of the commons' seems too compelling an addiction for us to escape. And no one seems able to break our tendency towards habits of maximising short-term gain despite their obviously unsustainable direc-tion and probable disastrous outcome.

The quota system and fish size limits, laudable though they may be in intent, also relate only to landed fish. It is commonplace for commercial fishermen to throw back netfuls of under-sized (dead) fish that they are not permitted to land, representing a complete waste from every conceivable angle. If we are serious about biodiversity conservation, we have to become hugely more intelligent in the way we apply theory into practical measures to deliver optimal outcomes for the greater and more enduring benefit of all.

We also fish progressively deeper waters, catching slow-growing fish that we may be driving to extinction before we have learned anything substantive about their ecology and potential for sustainable yields – that is, if the concept of 'sustainable yield' can practically apply at all to species or fisheries with such slow regeneration rates. And witness the madness of burning millions of tonnes of aviation fuel to fly fish across the world to serve remote, luxury markets. The basis of and benefits from a sustainable relationship with biodiversity is brought into sharp focus by the lessons we must learn from historic and continuing global fishery policies and their disastrous consequences.

3.2.6   Subsidies, fishing and biodiversity

Overcapacity in the world fishing fleet is a major cause of the depletion of the global fish stocks, with payment of government subsidies to the fisheries sector widely perceived as one of a number of practices responsible for over-fishing. Global fishing fleets are estimated to be up to 250% greater than what is needed to catch what the ocean can sustainably produce. One of the more surprising outcomes of the Fifth Ministerial meeting of the World Trade Organisation (WTO) in Doha, Qatar, during November 2001 was the commencement of negotiations that 'aim to clarify and improve WTO disciplines on fisheries subsidies, taking into account the importance of this sector to developing countries'. This agreement marks the first time where environmental conservation and

sustainable development have played a major role in the launch of a trade negotia-
tion, acknowledging the call by world leaders, formalised in 2002 at the World
Summit on Sustainable Development in Johannesburg, South Africa, for action
to maintain or restore world fish stocks to sustainable levels including the elimination of harmful subsides. In the same time period, the Food and Agriculture
Organisation of the United Nations (UN FAO) began to pay considerable attention to the role of subsidies adopting, in 1999, a (voluntary) International Plan of
Action (IPOA) on the Management of Fishing Capacity that calls on FAO members to reduce and progressively eliminate subsidies contributing to overcapacity.
Environmental NGOs including WWF and Greenpeace have been vocal in call-
ing for the WTO and national governments to address this issue [1, 2].

The scale of subsidies is large, albeit often less than transparent. Subsidies
take many forms, ranging from fuel subsidies favouring deep-sea trawling and
long-range fishing fleets, tax breaks for new vessels and payments for scrapping
old ones, direct income support, construction or development of port facilities
and even 'tied aid' to developing countries that buys the donor country access
to fishing grounds. The subsidies increase the capacity of the fleets of subsi-
dising nations, not only giving them a competitive advantage but also creating
excess fishing capacity that outweighs available resources. Not only do too many
boats chase too few fish, but wealthy nations also exhibit a disproportionate
capacity to exploit the fisheries of other nations who use less sophisticated and
expensive technology. Subsidised fishing may also drive interest in previously
unexploited or uneconomic fisheries including those further offshore or in the
fishing grounds of developing countries. Estimating the scale of fishing subsidy
is not easy, as much of it lacks transparency and public accountability. However,
various published estimates of global fishing subsidies range from $10 billion to
$15 billion annually, the largest in Japan and put at about $2 billion, all of which
equates to some 20–30% of annual trade in fish [3]. Of course, not all subsidies
are harmful, and it is important to differentiate those that are harmful from those
that are not and others that support sound environmental conservation (setting
up marine reserves and so forth), well-designed 'buy-back' schemes aimed at
decommissioning fishing vessels, or otherwise help eliminate overcapacity or
over-fishing.

Subsidies are far from being the only problems in world fishing. Management
failures of all types and scales play their part, although these may in effect be
exacerbated by subsidies. Lack of scrutiny and enforcement of fishing activities
obviously introduces further problems. However, most NGO observers, not to
mention the World Bank, the UN FAO and other influential intergovernmental
bodies, see subsidies playing a key role in either driving more or less sustainable
exploitation of marine biodiversity.

3.2.7   Hard lessons

The above discussion of the consequences of exceeding the 'carrying capacity' of
marine fisheries provides a set of generic lessons that apply to the exploitation of

other types of biodiversity stocks and ecosystems across the globe. We will apply them to positive solutions in Part IV.

References

[1] WWF, *Turning the Tide on Fishing Subsidies: Can the World Trade Organisation Play a Positive Role?* Gland, Switzerland: WWF, 2002.

[2] Allsopp, M., Page, R., Johnston, P. & Santillo, D., *State of the World's Oceans.* Guildford: Springer, 2009.

[3] Porter, G., *Estimating Overcapacity in the Global Fishing Fleet*, Gland, Switzerland: WWF, 1998.

3.3   Felling our future

Forestry provides another example to illustrate some of the adverse consequences of ill-advised over-exploitation of biodiversity for short-term gain. We'll have a brief look at some examples, from which we can extract a few important lessons.

3.3.1   Sweden's forest heritage

Until relatively recently, Sweden's industry was overwhelmingly forest-based. It is inherent in the culture of managed forestry that one plants and manages for the long term with stewardship as part of the ethos, unlike the expectation of instant or at least quick payback that dominates many other industries. However, around the end of the nineteenth and start of the twentieth centuries, Sweden faced two huge blows. First, over-harvesting of the seemingly boundless resource of trees had reached a point where the forests were in a severely depleted state, with drastic economic consequences. Second, at around the same time, the nation was hit by a devastating famine that claimed many lives.

So Sweden knows from relatively recent cultural memory that adversity born of ecosystem depletion can have serious consequences for human well-being and the economy. It has also, within recent folk memory, had to cope with radical reform of policies to place sustainability, for the long-term well-being of both society and the businesses that serve it, very much to the fore. This in my view explains much about the sustainability ethos and environmental leadership demonstrated in Sweden, which should give us some hope that it is possible to challenge and change public policy elsewhere in the world.

3.3.2   Not just the trees

However, it is not merely the long-term security of wood that is at issue when forest ecosystems are removed. Wood may be an economically valuable asset to harvest, yet deforestation can be costly in other ways.

Where extensive forest clearance occurs, the hydrology of whole watersheds can be radically altered. For example, the unprecedented flooding in the Yangtze River basin during the summer of 1998 drove 120 million people from their homes and, although initially referred to as a 'natural disaster', it was later realised that the removal of 85% of the original tree cover in the basin had left little vegetative cover to hold the heavy rainfall. Not only was direct flood damage consequent from forest removal, but also the surge of storm water, no longer slowed and stored by mature forests, carried with it huge quantities of topsoil, depleting the primary resource of agriculture.

The same principle applies in peninsular India, where the forested mountain chain of the Western Ghats that borders the western edge of the subcontinent intercepts moist south-west monsoon winds blowing in from the Arabian Sea. The rainfall here is seasonally significant, the Western Ghats consequently being

naturally vegetated by rainforest, dry and cooler upland forests and more open savannah-like shola forests, which are not only ecologically important but also perform a range of ecosystem functions of crucial importance to the whole of Deccan India, significantly including the trapping and retention of moisture. Eastwards beyond the Western Ghats, the Deccan peninsula is increasingly arid yet is moistened by the three vast river systems draining the Ghats – the Godavari, Krishna and Cauvery – spanning the entire continent, and conveying life-giving waters across it. Nevertheless, commercial forestry, resort development and rainforest clearance on these functionally vital mountains is believed now to be compromising their capacity to capture and store rainfall, which has ramifications throughout the whole socio-ecological river system including for the millions that make their livelihoods there. A major concern is the threat of biodiversity loss for security for the already over-stretched water supply to the whole of peninsular India.

The same principles also apply to broad-scale forest felling in the Amazon, South Africa, Kenya, Indonesia and many other parts of the world. I will not labour the point any further here. However, let's remind ourselves that these widespread examples reinforce the message that forests produce a huge diversity of ecosystem services beyond the production of wood. The pressures we place on forest products have wider environmental and economic ramifications as well as impacts on the vulnerable societies reliant on the functions of forests and the catchments in which they occur.

As for marine fisheries, stemming the flow of short-term profit-taking from over-exploitation of forests is hard to achieve even where the economic and societal costs have been demonstrated. The massive extent of clear-felling, both legal and illegal, of Amazonian and Indonesian rainforests is well publicised, and also highlights how loss of ecosystems can result in hardship for forest-dependent human societies as well as the wider loss of species, depletion of biological resources that may be important for pharmaceutical purposes, remobilisation of pollutants and a wide range of other human and ecological consequences.

3.3.3   Subsidies, forests and biodiversity

Industrial forest plantation subsidies obviously introduce a range of impacts and implications for biodiversity and other aspects of global forests. Some of these subsidies may again be hidden, such as the construction of road and rail access into areas of old-growth forest, which stimulates exploitation by legal and illegal means. Income support and grants for forestry equipment are a more direct form of subsidy that may not work to the advantage of forest ecosystems.

However, not all subsidies need be negative for biodiversity. Significant old-growth forest has been lost around the world, though official figures in regions such as the UK, Europe and India show recovering areas of forest cover largely due to more recent plantations. This highlights a need for protection of the remaining resource particularly where it serves important ecosystem functions such as water capture and storage or climate regulation. Subsidies may create an incentive to favour this type of outcome.

In managed forest systems, biodiversity conservation preferably requires both prolonged rotation age and retention of some stands of trees and other residual habitat on felling, ideally also leaving sufficient decaying and dead wood to sustain a variety of species dependent on it. Given that this works contrary to economic pressures to capitalise on the felling of all trees in a forest, subsidy and tax instruments can play a significant role in enhancing biodiversity in managed forests too. This approach is being applied in many European countries, including subsidies for multiple-use community forests explicitly intending to protect or enhance biodiversity. In the US, some subsidies address the linkage between landowner and regional impacts, noting the benefits to the 'common good' that can ensue through prudent or restrained management of private forest property.

Elsewhere, particularly where tracts of old-growth forestry remain, subsidised eco-forestry, including alternative beneficial uses of forest such as low-frequency cropping, tourism and so forth, or indeed strictly enforced limits on the extent of timber extraction, may produce an economic return to landowners comparable to that realised by industrial logging.

3.3.4   Making the business case work

As another example, a cost–benefit case study involving Madagascan rainforests revealed that their conservation would generate significant benefits over logging and agriculture, both locally and globally. However, despite a robust business case for protection addressing the clear and enduring benefits that this would yield, logging continued, driven by greater incentives at the national scale. Such perverse economic 'signals' continue to exacerbate tropical deforestation and wetland loss worldwide, even where we can already foresee the long-term harm that will ensue.

This principle of perverse economic beneficiaries applies equally in the management of river catchments, wetlands and coastal zones, where most of the benefits of mismanagement accrue to minority interests who may not pay for the actions that deliver them and their ensuing consequences. In the context of coastal woodlands, clearing of mangroves from India, Sri Lanka, countries around the Gulf of Mexico and many other tropical shorelines, generally for short-term gain such as for charcoal production, shrimp farming, tourist resorts or sugar plantations, has been seen as a major factor increasing the extent of storm damage through loss of the buffering effect of vegetation. Furthermore, removal of coastal forestry can accelerate land erosion, the loss of the precious resource of soil, siltation of lower rivers and smothering of coastal fisheries and sensitive ecosystems.

Biodiversity matters, and generally for a broader suite of reasons than is commonly appreciated. More sophisticated economic and legislative tools are essential in forests and other ecosystems if sustainable management is to be realised, including linking the beneficiaries with those in a position to conserve biodiversity resources with all of their associated functions.

3.4   Breaking ground

The topic of land use for agriculture is a vast one, and also an ancient one stemming back over probably 8,000 years or possibly more. Indeed, Davies and Day make the grand statement in their 1998 book *Vanishing Waters* that 'Modern irrigation-driven monocultural agriculture is probably the single most environmentally devastating development in the entire history of our species' [1]. They back this up with the point that there has been a profound alteration to a huge extent of land across the globe, converted from the state in which humans first found it, and vastly different in appearance, biodiversity and function than more 'natural' habitats such as those seen in the world's better nature reserves.

3.4.1   Changing the landscape

Agriculture, encompassing everything from arable crop production, grazing, orchards, vineyards and intensive livestock rearing, produces the overwhelming bulk of the food we eat, as well as feedstock for livestock and natural fibre. Increasingly, crops are being used as a source of biomass-derived energy, chemical feedstock, pharmaceuticals, dyes and other commodities. Agriculture is central to human history, with evolving agricultural methods enabling socio-economic metamorphosis across the world. Until the Industrial Revolution, agriculture dominated the economy of Europe, and it still does so today in non-industrialised countries where it supports basic needs and provides tradable goods. Currently, an estimated 42% of the world's workers are employed in agriculture, by far the most common occupation on the planet, although agricultural production accounts for less than 5% of the gross world economic product.

With such a massive extent of agricultural activity and a diversity of methods, Davies and Day's statement does not appear unreasonable. And, when we look at the huge levels of energy, agrochemicals and water used in modern intensive agricultural production methods, the implications for biodiversity are equally massive. Indeed, cropland and range lands as well as forests are among the more pressurised habitats across the entire planet, as determined under the Millennium Ecosystem Assessment. Sharp downward trends in farmland wild bird populations in the UK since the middle 1970s provide a stark illustration of the negative impacts of contemporary industrialised farming methods on biodiversity [2]. The trend has now stabilised somewhat, though there is still a slow decline in species with specific adaptations to these habitats, offset largely by a rise in population of more generalist species.

This chapter is, however, only an overview and so I will leave it to such well-researched sources as Jules Pretty's 2002 book *Agri-Culture* to provide additional pertinent facts and figures for those with further interests [3].

3.4.2 Subsidies, land use and biodiversity

Of the wide range of market-based incentives available to governments, subsidies are a favoured instrument for biodiversity protection, offering considerable potential for promoting more biodiversity-friendly practice in farming and other fields governed by private interests. However, agricultural subsidies are far from straightforward in their effects. In practice, subsidies for changing land use or management may, for example, result in inadvertent consequences that eliminate other intended benefits, including those of other subsidies applied elsewhere for the purpose of nature conservation. This sort of 'perverse subsidy', delivering unintended negative outcomes for biodiversity and public well-being, is best exemplified by European agricultural policy in the last four decades of the twentieth century. Another example concerns agricultural fertilisers that contribute to marine eutrophication including the formation of hypoxic 'dead zones'. The Greenpeace report *State of the World's Oceans* attributes partial recovery of one such dead zone formed in the north-western Black Sea in the 1970s and 1980s, at one time covering up to 40,000 km^2, with some certainty to the reduction in use of agricultural fertilisers resulting from the economic collapse of the former Soviet Union and declines elsewhere in subsidies for fertilisers [4]. Government subsidies for ecologically unsound aquaculture practices are also believed to have contributed to diminished protection of coastal ecosystems such as wetlands and mangroves, increasing the consequences and costs of storm damage as well as causing wider damage to biodiversity.

Even agri-environment subsidies, designed to lessen the impacts of farming on biodiversity or indeed to create habitat to encourage it, are far from straightforward in outcome. (They have been already partially addressed earlier in this book.) In general, academic studies into the efficacy of agri-environment schemes for biodiversity conservation have been equivocal, demonstrating highly variable effectiveness in delivering biodiversity benefits and indeed producing disproportionate gains or losses for different taxonomic groups. Often, the subsidies are imprecisely targeted, thwarted by lack of uptake, or monitored too poorly, if at all, to assess effectiveness and scope for improvement. A report to the EU by Ecologic (Bäuer *et al.* [5]) in 2005, informed by numerous case studies worldwide, identified the need for various criteria to make market-based instruments more effective. These include clear objectives, definition of the good to be traded, social effects, unexpected environmental effects, pilot-testing, setting up on an appropriate timescale, flexibility in implementation and subsequent monitoring and auditing of effects. There is no doubt that the existing agri-environment schemes could be improved by such measures.

3.4.3 Healing the land

Like all contemporary economic activities in the consumer economy, capitalisation of assets such as productive land, mined nutrients and water sources have been effective in generating short-term income from agriculture in ways largely

ignorant of their long-term consequences for biodiversity and wider ecosystems. I will also touch on some more examples in the following chapter dealing with water. The clear lesson now, knowing what we do and reflecting on its implications both for humanity and also for business depending on agriculturally derived products, is that nothing less than a revolution is required to embed sustainability principles into future production methods.

References

[1] Davies, B. & Day, J., *Vanishing Waters*, University of Cape Town Press: Cape Town, 1998.
[2] Defra (UK Department for Environment, Food and Rural Affairs), *Biodiversity Indicators in your Pocket*, Defra: London, 2007 (http://www.defra.gov.uk/news/latest/2007/biodiversity-0612.htm, accessed August 2007).
[3] Pretty, J., *Agri-Culture: Reconnecting People, Land and Nature*, Earthscan Publications: London, 2002.
[4] Allsopp, M., Page, R., Johnston, P. & Santillo, D., *State of the World's Oceans*. Guildford: Springer, 2009.
[5] Bäuer, I., Müssner, R., Marsden, K., Oosterhuis, M., Rayment, M., Miller, C. & Dodoková, A., *The Use of Market Incentives to Preserve Biodiversity: The Project under the Framework Contract for Economic Analysis ENV.G.1/FRA/2004/0081*, Ecologic: Germany, 2005.

3.5   Liquid assets

The story of water resources across the world offers further salutary warnings for the direction of human development. The following examples are drawn from a wide suite of research but some particularly useful facts were published in Lester Brown's 2001 book *Eco-economy: Building an Economy for the Earth* [1].

3.5.1   A medium for development

Water is a resource we continue to take largely for granted, despite the fact that it has long been acknowledged as a limiting factor to human well-being and development worldwide. Indeed, explicit commitments to providing water resources for the developed world are part of the United Nation's Millennium Development Goals. Among the resolutions of the UN World Summit on Sustainable Development (WSSD) in Johannesburg, 2002, was a target to halve the number of people without access to adequate sanitation (now 2.4 billion people) by 2015. 2003 was designated by the United Nations as the 'International Year of Freshwater', the aim of which was to focus attention on protecting and respecting our water resources, as individuals, communities and countries.

Yet, as living standards rise along with material expectations, concerns about water quantity and quality are set to increase. These are, of course, exacerbated by the impacts of climate change. In some areas, especially cities, rapidly growing populations are making demands on water far in excess of available supplies. Even when there is sufficient water, distribution infrastructure can be woefully inadequate leading to inequities in access. An estimated 26 countries with a combined population of more than 300 million people suffered from water scarcity in 2001. By 2050, projections suggest that 66 countries, comprising two-thirds of the world's population, will face moderate to severe water scarcity. Today, about 1.2 billion people, a fifth of the world's population, do not have access to clean drinking water, while 2.9 billion people lack access to adequate sanitation facilities. These factors combine to drive them downwards in a spiral of increasing disease, lost productivity and potential and higher costs entailed in obtaining water.

3.5.2   Running into the sand

Some of the world's major rivers are being drained dry, failing to reach the sea, and not all of them are in remote corners of the globe. One example is the Colorado River, which drains a part of the arid regions on the western slope of the Rocky Mountains and constitutes one of the major rivers of the south-western United States and north-western Mexico. Naturally, the river would discharge into the Gulf of California, but heavy use of water for irrigation has desiccated its lower course in Mexico. Today, the Colorado no longer consistently reaches the sea. A series of large dams on the upper river and throughout its entire course compound this situation. Yet, before the mid-twentieth century, the Colorado River Delta comprised a rich estuarine marshland.

In China, the Yellow River, the northernmost of the country's two major rivers, no longer reaches the sea for part of each year. In Central Asia, the Amu Darya sometimes fails to reach the Aral Sea because it has been drained dry by upstream irrigation demands. The Aral Sea itself is shrinking beneath the relentless sun in this semi-arid region. Since 1960, the sea has dropped 12 m, and its area has shrunk by 40% and its volume by 66%. Towns that were once coastal are now 50 km from the water. If recent trends continue, the sea will largely disappear within another decade or two, becoming a geographic memory existing only on old maps.

At a wider geographical scale, there is a threat to the balance of water fluxes between continents and oceans. This is, of course, exacerbated by human-wrought disturbance of the climate.

3.5.3   Wetlands and groundwater

I have already addressed some of the interactions between land cover and the water cycle when considering forestry, and it is clearly pertinent to agriculture as well, particularly through irrigation and changes to catchment hydrology. Natural ecosystems, especially wetlands and forests, capture water and stabilise seasonal flows, while recharging groundwater and improving water quality. They are also hotspots for both biodiversity and natural productivity, the biodiversity doing the work of water purification and storage. Conserving wetland ecosystems is vital in maintaining the supply of renewable fresh water, yet half the world's wetlands were lost to development during the last century.

And, as we over-harvest the scarce resource of groundwater, we also pollute it. UNESCO estimates that there is a $5 billion market in remediating contaminated groundwater and land throughout Europe, reflecting the compromise of critical resources and the greater hidden costs arising from issues 'out of sight and out of mind'. Pollution from agriculture, industrial and municipal sewage, in addition to salinisation from irrigation, has also reduced the availability of clean fresh water.

The world is also running up a water deficit as rising development pressures across the world lead to both surface waters and aquifers running dry. Irrigation problems are as old as irrigation itself, but the new threat that has evolved over the last half-century with the advent of powerful diesel and electrically driven pumps dwarfs yesterday's risks. The over-pumping of aquifers, now commonplace on every continent, has led to falling water tables as pumping exceeds natural recharge rates. Water reserves are being used up or polluted faster than they can be replenished. Around 1.5 billion people today rely on groundwater for drinking, and 10% of people's water consumption worldwide is from depleting groundwater.

Water tables are falling over large expanses of all key food-producing countries. Under the North China Plain, which accounts for 25% of China's grain harvest, the water table is falling by roughly 1.5 m per year, and probably much faster than that in some places. The same thing is happening under much of India, particularly the Punjab that is the country's breadbasket. In the United States, water tables are falling under the grain-growing states of the southern Great Plains, shrinking the irrigated area. Today, 480 million of the world's 6.1 billion people are being fed with grain produced by the over-pumping of aquifers. The long-term implications for the sustainability of the aquifers, together with the potential for food production, are ominous.

3.5.4   Water and crops

We live in a water-challenged world. And it is becoming more so each year as 80 million additional people stake their claims on the Earth's water resources. The situation promises to become far more precarious, since most of the 3 billion people added to world population by 2050 will be born in countries already facing water scarcity, and quickly moving into water poverty. With 40% of the world's food supply coming from irrigated land, water scarcity directly affects food security. If we are facing a future of water scarcity, we are also facing a future of food scarcity and one that is increasingly threatening for wildlife. Food supplies may therefore be a vulnerable link between environment and economy. An estimated 70% of the water consumed worldwide, including that diverted from rivers and pumped from underground, is used for irrigation. Of the remainder, 20% is used by industry and 10% for residential purposes. In the increasingly intense competition for water among these three sectors, the economics of water do not favour agriculture. Neither do they favour the continued existence of viable sources of water, the ecosystems that depend on them, nor long-term human well-being dependent on these ecosystems.

3.5.5   Hydrological poverty

Hydrological poverty – lack of access to adequate safe water for basic needs and economic activities – is a local form of impoverishment that is difficult to escape. It will become more common in coming years. This is of particular concern in arid areas, and where there is competition for severely limited resources. For example if the Nile basin countries do not quickly stabilise their populations, they risk becoming trapped in inescapable hydrological poverty with all of its attendant risks of disease, human misery, constraints on economic and social progress, lack of opportunity and potential for conflict.

South Africa's far-sighted National Water Act of 1998 has addressed empowerment of 'Historically Disadvantaged Individuals' as a priority alongside the ring-fencing of an 'Ecological Reserve' to support wildlife, support for basic human needs for all people, and additional reserves for national security (comprising mainly energy security) and international commitments to sustain river flows into neighbouring countries. These take precedence over other uses of the water produced by river catchments across the semi-arid country, the remainder then being available for allocation to other users. The water resources of South Africa are already significantly over-stretched, even before addressing such new priorities as allocating water to emerging black farmers and stimulation of a 'second economy' among those historically disadvantaged. Therefore, all of South African society has to collaborate and to innovate together to ensure that the National Water Act's key objective of 'equity, sustainability and efficiency' in water use is met without destroying the ecosystems on which long-term security of water depends. The prioritisation of a 'non-negotiable' Ecological Reserve, reflecting the needs of the ecosystem, is a welcome regulatory step, although there are clearly regional and local concerns about how that reserve is calculated and apportioned across catchments and hydrological regions.

Although the UK does not face the kinds of severe water problem experienced overseas, there is no room for complacency. Pressures such as new residential and industrial development, climate change and excessive domestic consumption are substantial and increasing, particularly in the south-east of the country where the water resources are already most significantly over-stretched. In May 2006, the UK government issued drought orders earlier than ever in a year following 19 months of below-average rainfall, amounting to the worst water shortage in recorded history. The news media were quick to blame the (privatised) water industry for leakage from pipes and other forms of wastage, which certainly are issues that need to be addressed. However, a significant part of the larger picture is the spiralling demand of more people with water-hungry lifestyles, the assumption by some of a fundamental right to unlimited access, and overcrowding of population in the driest quarter of the country. To address sustainable water supply strategically, everyone has to learn to see themselves as part of both the problem and the solution.

Looking ahead to the probable range of impacts of a changing climate, it is unlikely that this situation will miraculously correct itself. Collectively, we have to take responsibility and find a sustainable relationship with water resources and their ecosystems. If our thirsty habits degrade the critical ecosystems within catchments, we may lose more of the hydrological, water purification and other water-related functions that they perform, compounding the cycle of degraded biodiversity with adverse consequences for humans.

3.5.6   Biological wealth drying up

Not only is aquatic and wetland life threatened by our over-dependence on water but, by changing the nature of ecosystems and their functions, the capacity of biota to sustain human well-being, economic activities and enjoyment is also prejudiced. And, as for any other aspect of biodiversity, we dry up our wet wealth and the life it supports at our collective peril.

Reference

[1]   Brown, L.R., *Eco-economy: Building an Economy for the Earth*, W.W. Norton: New York, 2001.

3.6　Biodiversity in the balance

I am wary of confronting the reader with a saga of ecological catastrophe and disempowering despair, but it is nevertheless important to convey some measure of the extent to which historic and continuing exploitation of the planet's biodiversity is having serious impacts. It is also important to illustrate that this has serious consequences for human well-being and corporate profitability, and to provide a perspective for the nature and proportion of response that is due. The biological diversity of our home planet does, after all provide a direct 'read-out' of society's impacts. The messages arising from this analysis are far from gleeful.

3.6.1　The thickening red line

With each update of its *Red List of Threatened Species*, the World Conservation Union (IUCN) charts an increase in all categories of 'critically endangered' species. The 2000 IUCN assessment revealed that 1 in 8 of the world's bird species is in danger of extinction, about 70% of bird species are declining in numbers, 10 (out of 17) species of penguins are threatened or endangered and a quarter of the mammals and one-third of fish species face extinction. The 'bush meat' trade together with progressive habitat loss exacerbates the situation for chimpanzee species, as for example in West and Central Africa where 97% of bonobos (the chimp species most closely related to humans) have disappeared in less than a human generation.

3.6.2　Mixing it up

A major, yet commonly underestimated, threat to biodiversity of all types is the introduction of alien species, which can alter local habitats and communities and introduce exotic diseases, driving native species to extinction. Although this phenomenon is as old as human adventure – for example the impacts of rats introduced into island ecosystems or the deliberate introduction of European foxes and domestic cats to Australia – the pace of species transfers around the world seems to be quickening dramatically.

One consequence of globalisation, with its expanding international travel and commerce, is that more and more species are being accidentally or intentionally brought into new areas, where natural controls are absent, and are thriving to the detriment of established ecosystems. Of all the bird and plant species on the IUCN Red List, 30 and 15% respectively are threatened by non-native species. The extent of the threat posed by the coincidental introduction of novel diseases is substantial but harder to quantify.

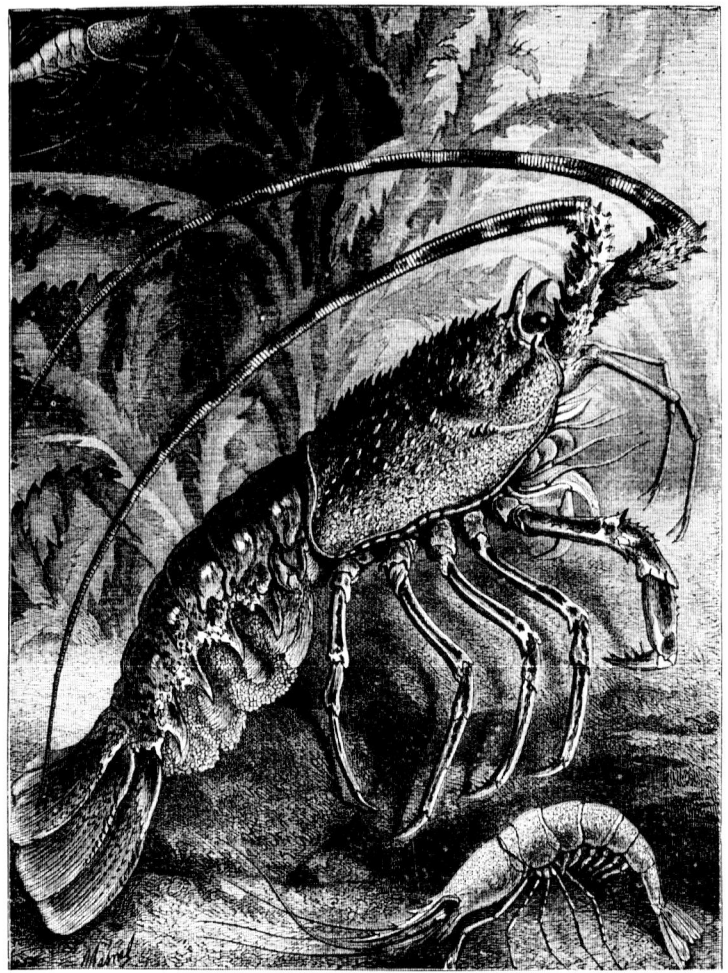

3.6.3   Declining diversity

No one knows how many plant and animal species there are on Earth today, though current estimates range from 6 to 20 million species with the best working estimates falling between 13 and 14 million. It is impossible to fully quantify the biological effects of the most recent explosion in human economic activities since our knowledge of the number of species, and their global and local distribution, is so incomplete.

We can measure losses where we have something approaching a substantially complete inventory of species, such as for birds. However, for the many groups of small organisms, whose species number in the millions (such as insects), only a fraction have been identified, described and catalogued. We understand far less about their roles within ecosystems, their potential as a source of pharmaceuticals or other resources, and the consequences of their loss.

We know virtually nothing about microbial biodiversity and function, beyond the fact that it is precisely these 'bugs' that are the major players in the functioning of global chemical cycles. Without them, the visible life forms with which we are most familiar, including ourselves, simply could not survive.

3.6.4   Declining functionality

As has been repeatedly emphasised in these pages, biodiversity is important not only for what it is but particularly for what it does. The preceding chapters in Part III and those in the parts to follow outline how the functions of river catchments, forests and other ecosystems produce a diversity of goods and services that support the needs and aspirations of society now and into the future. Definitive and unambiguous methods for evaluating ecosystem functionality remain elusive, although some indicative estimates are provided with respect to water resources elsewhere in this book.

There is no shortage of examples of what biodiversity loss means in practice. Collapsing marine fisheries may result in direct employment losses in the thousands, indirect job losses in the tens of thousands, devastation of local communities and financial costs running into billions of dollars, and the loss of ways of life and vital ecosystems. Declining soil fertility, the loss of pollinators and the disturbance of ecosystems and their functions through the spread of invasive species are more difficult to quantify. Studies have suggested with confidence that the destructive impact of the hurricane that hit New Orleans in the summer of 2005 would have been much less severe if coastal wetlands, which naturally absorb a great deal of coastal storm energy, had not been so extensively degraded or destroyed. In a similar vein, the impact of the Asian tsunami in December 2004 would have been significantly reduced if the buffering effect provided by coastal mangroves had not become fragmented through human activities.

Nature underpins our economies and other life expectations and aspirations, and the functions that nature performs need to be recognised, respected, valued and protected.

3.6.5   Sliding to extinction

Best estimates suggest a current species extinction rate at least 1,000 times higher than the background rate (i.e. without human interference). Irreplaceable species and ecosystem services, the products of 3.85 billion years of evolutionary fine-tuning, are being expunged by the march of technological progress over infinitesimally small timescales. As its diversity diminishes, nature's capacity to support the growing needs of humanity shrinks. Nature's larder, pharmacy, supply cupboard and cleansing service is depleted, depriving future generations of new discoveries, potential for sustenance, economic opportunity, heritage and 'quality of life'.

I need not repeat here the evidence provided by the Millennium Ecosystem Assessment, outlined in the chapter 'Biodiversity and sustainable development'. Suffice to say that rigorous scientific evidence supports the view that global ecosystems are severely depleted and that the pace of degradation is accelerating.

Biodiversity is an essential resource for our own survival – economic as much as physical – yet we seem unable to stop ourselves killing this metaphorical 'goose that lays the golden egg'.

3.7   Changing course

Newspapers and other media are littered with instances of what happens when businesses get caught out harming biodiversity, whether inadvertently or wilfully. I'll not go into any great depth on this matter: just a small smorgasbord of examples will be adequate to illustrate the implications for corporate reputation when businesses fail to account for and respect biodiversity and associated public opinion.

3.7.1   Biodiversity and reputation

It was at 12:04 am on 24 March 1989 that the giant tanker ship *Exxon Valdez* ran aground, dumping over a quarter of a million barrels of oil into Alaska's Prince William Sound. Still the largest recorded marine oil spill, damage to the fragile coastal ecosystem was severe including tens of thousands of seabirds and otters, hundreds of bald eagles, many whales, thousands or tens of thousands of fish and untold consequences for invertebrates, seaweed and other less prominent marine life forms. Deformities in newly hatched fish and other wildlife are still reported as a consequence of persistent pollution. The human cost in terms of impacts on fishing communities were initially substantial, and is also long lasting as impacts on the ecosystem will endure for generations. The manner in which the spill was handled by the tanker's owner, the oil giant Exxon, attracted widespread condemnation by environmentalists and the media alike in terms of the scale, pace, effectiveness and end-point of the response. The immediate economic impact for Exxon was a reported 5% sales loss. However, key *Exxon Valdez* into an internet search engine today and the strength of feeling towards the company elicited by this event (bolstered by several others besides) clearly continues to blight its reputation. Though the impact is hard to quantify in economic terms, the fact that many organisations still treat Exxon as a pariah is clearly not to its economic advantage. The public, which includes individuals that influence corporate customers, cares and is out there watching. Once breached, corporate trust can take a very long time to recover.

For some people, whatever the realities and subsequent changes in the performance of the company, the name 'McDonald's' will ever be linked with the clear-felling of rainforest areas in South America for creation of short-lived grazing for beef cattle production. This was amplified in the public psyche by the long-running and high-profile 'McLibel' trial. Various commentators and activists, particularly throughout the 1990s, criticised McDonald's global promotion of beef consumption despite the reputed harm inflicted by cattle ranching on tropical forests. McDonald's refuted these claims, stating that the company does not use beef from cattle reared on recently deforested land and that its written rainforest policy statement of 1989 merely formalised a verbal policy in place since McDonald's opened its first store in 1955. However, defendants in the trial argued that this was blatantly untrue, providing evidence to support their case. Whatever the rights and wrongs, and irrespective of the final decision of the court, the reputation has stuck to McDonald's. It highlights the sensitivity of the public to exploitation of biodiversity and of people dependent on it for traditional lifestyles, both of which can offend deeply held values. As a major American-based multinational brand, exporting what some refer to as 'McCulture' around the world, McDonald's has also since been targeted by anti-capitalist demonstrators in France, Switzerland, Germany, the UK and many other countries besides. The perception of big corporate interests riding roughshod over local interests and ecosystems is a major part of the issue to which the demonstrators object, with McDonald's perceived as an icon of the so-called 'Coca-colonisation' of the world by American brands and their underpinning commercial practices. Certainly, McDonald's is a soft target due to its size, global reach and instantly recognisable branding. However, whether or not our own enterprises are anything close to the scale of such global businesses, we should be forewarned about dismissing too readily the reputation issues that can attach themselves to an organisation for perceived impacts on biodiversity and the local livelihoods that depend on it.

Calls for relaxation of restrictions on prospecting for oil in the Alaskan wilderness have been headline news throughout much of the US Presidency of George W. Bush, who overturned wildlife and landscape conservation measures established by the former Clinton regime. For many Americans, the issue was a battle between the supremacy of commerce over all other concerns including the value of wilderness for its own sake, with all the ecological, spiritual, recreational and 'First Nations' issues that attach to it. The ecosystem of much of coastal Alaska is indeed remarkable, and based around the lemming cycle. When lemmings are plentiful, predators such as jaegers, owls, foxes, wolves, weasels and wolverines prosper and, when the lemmings are scarce, predator numbers also fall in response. This reflects a beautifully adapted balance evolved in an inhospitable climate. Upset this balance, however, and the fragile ecosystem can no longer cope. In 1906, the United States Geological Survey discovered oil in the coastal area near Prudhoe Bay; by the end of the decade, oil companies had begun staking claims in Alaska's North Slope. For some commentators, America's greed for oil has drastically upset the ecological balance, and is spreading to consume increasingly large tracts of Alaska besides. And, as prospecting has revealed more rich oil-bearing deposits across the formerly protected wilderness of Alaska, oil companies are clamouring for a stake in the profits to be made in servicing the hunger for energy that underpins the American way of life. Politically, ecosystems and their champions seem powerless, often perceived as merely standing in the way of realisation of the 'American Dream'. Although company names are currently rarely linked with this exploration and ensuing exploitation, it is certain that, if it is to follow the pattern elsewhere in the world, we can expect public 'naming and shaming' and boycotts of businesses as people feel that nature falls victim to corporate greed. Reputation and competitive edge will again suffer, as has increasingly become the case when business practice offends public values with respect to biodiversity.

Monsanto is another company not greatly loved by the environmentally conscious. The scale of public disapprobation and the breadth of reasons for it are also quickly revealed by keying in the company name on an internet search engine. One of the key emotive terms attached to the company is that it has been 'playing God' or 'patenting life' in its promotion of genetically modified technologies. Faced with fish genes implanted in tomatoes, 'terminator' genes being

inserted into crop species to kill them before they can set viable seeds, micro-bial insecticides introduced into plant genomes or other interspecies transfers of genetic matter that override natural checks and balances, people become deeply uneasy. They become actively antagonistic when they feel that this technology is being forced on them, or on farmers in less developed nations who may be forced into dependence on agri-business multinationals rather than being able to profit-ably pursue a traditional form of agriculture. The reputation issues for Monsanto stemming from the perception of tampering with life itself has earned it a pariah status with many groups and people, and set back serious research into and debate about the potential of genetic modification technologies.

We could look at the implications for the Spanish strawberry-growing industry from revelations of substantial wetland loss through over-abstraction of ground-water for irrigation, or the reputation issues attaching themselves to fish kills from industrial and agricultural accidents or malpractice. Or we could summa-rise the fate of the fur trade and the cosmetics industry when animal rights issues come to public attention. However, the point has probably been adequately made already. Damage to reputation, with its raft of market, investor, staff retention and corporate morale implications, is a near-inevitable consequence when businesses are ignorant or negligent about harm to biodiversity.

3.7.2   Can business draw us back from the edge?

The truth of the matter is that the hunger of a growing population for a dwin-dling base of natural resources, as yet barely constrained by national regulation or international agreements, is currently way in excess of ecological 'carrying capacity'. We are, in many ways, teetering on the edge of disaster as the very biological systems on which society depends are progressively over-exploited beyond their sustainable limits.

As we have seen in the case of fisheries and forests, it seems that the availabil-ity of expert scientific knowledge and foresight alone has to date been inadequate to arrest established cultural momentum, vested interests and a blindness born of arrogance or ignorance. Yet, without drastic and prompt change in society's behaviour, we can only plummet into the abyss of collapsing ecosystems and their associated supportive functions.

The various examples provided throughout Part III if this book paint a some-what bleak yet wholly honest reality. And yet for every negative there is a positive alternative. It should be evident from these examples that, were businesses to have taken a more positive approach to biodiversity, they could have benefited from doing so in alternative ways.

We have to change as a society if we want a healthy, wealthy and fulfilled future. Even pessimists have to admit this is possible, and all of us together need to make that hope a practical reality rather than sleepwalking to disaster. Socie-ties can change with respect to their relationships with supportive ecosystems, as

Sweden has demonstrated, and business may just be a major leader and beneficiary of this 'change for the good'.

So enough of the negative stories! In the Part IV, we will look at how businesses can respond to the challenge of a responsible relationship with biodiversity, illustrated with some encouraging examples of how it can prosper when it does 'the right thing'.

Part IV Acting for biodiversity

4.1 So what can I do?

Biodiversity matters. Not just for altruistic reasons; it really matters. Humans are part of the biodiversity of this world, and will prosper or suffer collectively with the planetary web of life. This applies personally, culturally, regionally, nationally, globally and, of particular emphasis within this book, corporately too. If we erode the capacity of the world to support our diverse needs and activities, we can only serve to limit our prospects and potential. So, then, what can we do positively to act for biodiversity? And, critically, how can we do so in ways that create economic advantage for our business?

Fortunately, quite a lot. Some first steps are easy and offer quick payback, but many others will require further investment and the payback will not be so rapid. However, in a fast-changing world shaped by sustainability pressures, we can be assured that preparing ourselves to live and trade with a lighter tread on the Earth's fabric can only be beneficial in the long run. The pace at which we innovate and invest, then, is purely a matter of sequencing in successive business planning cycles, rather than a question of whether or not we should do so.

The following four chapters address, in brief outline only due to the constraints of space, actions appropriate to the protection of biodiversity. The first three of these following chapters address measures appropriate to businesses, or activities within business, at successive remove from the gritty interface with biodiversity. Each of these three is concluded by a section titled 'So what can my business do?' which presents a stepwise set of issues that can be used to question your business model and make initial progress with deepening a commitment towards your biodiversity dependencies. Chapter 4.4 addresses how emerging awareness of the need to protect biodiversity can generate new opportunities for business.

Actions offered in the following four chapters have self-benefit in mind, as business has to make incremental progress on a profitable basis. And, although they are necessarily only a quick skim through many possibilities, the ideas presented should give you the gist of how to address the relationship of your business with biodiversity on a responsible and ideally profitable basis.

In Part V, I'll synthesise all of these good ideas, supported by practical examples, into an integrated framework to progressively bring biodiversity concerns into the heart of business decision-making.

4.2   Business close to biodiversity

There are many types of business activities operating close to the interface between the economic and ecological strands of sustainable development, converting the latter into the former to deliver societal needs. These range from the water industry to forestry enterprises, various uses of land including production of food and biofuels, commercial fishery interests that harvest from nature and recreational fisheries that utilise fishery ecosystems and the natural beauty that surrounds them. All are, in essence, extractive industries that exploit populations of organisms and the ecosystems of which they are part.

For this category of business close to biodiversity, the need for an interdependent accommodation between human exploitation and ecosystem needs is often reasonably self-evident. Or, at least, it becomes so when the limits of carrying capacity are exceeded, whether through greed or ignorance, with inevitable harm for ecosystems and people. Practical examples of this have been provided throughout Part III, which took a brief trip through the doom and gloom of the consequences of getting that balance with biodiversity spectacularly wrong, with all the ecological, social and economic damage that inevitably ensures. In Part IV, I'll instead take a trek through the sunlit uplands populated by examples of businesses and other organisations which have got the balance right, often brilliantly so. There are some genuinely inspiring stories.

For the remainder of this chapter, I'll look at examples of water, forestry, fisheries, endangered species and food production, before then pausing to draw out generic lessons for business activities close to the primary resource of biodiversity.

4.2.1   Drop wise

Water is part of one of the great natural cycles of this planet. It circulates seamlessly throughout the biosphere, exchanging freely between land, air and water bodies. Evaporated water rises from the surface of the sea, fresh waters, wetlands, plants and soil, and this is then distributed and condensed in the atmosphere. Subsequently, it falls back as precipitation, to be stored in ice, snow or groundwater. Some re-evaporates, with ecosystems recapturing a variable proportion again in local water cycles that can keep a forested gorge green in an otherwise arid landscape. Water flowing through the landscape is subject to purification processes within aquatic habitats, before flowing from land to freshwater systems and, often, to sea or open fresh waters where it is retained for varying periods and may be evaporated once again. This cycle is perpetual, powered by solar radiation, providing an endless, purified resource that sustains myriad life forms for all or part of their life histories. Humans depend completely on this endlessly renewable water cycle. However, the way we use water often compromises the very resource on which we depend.

The concept of the catchment (or 'watershed' in technically inaccurate American parlance) protection zone is now well established across the world, particularly in the United States, Australia and the UK. In these territories, land has been enclosed or brought into public ownership or management explicitly for its capacity to intercept, store and purify precipitation. Still wider areas have been subject to management plans, often backed up by public subsidies, to prevent compromising this important function. Lake and river catchment planning recognises the critical importance of high-quality water resources, which also support biological resources that may further enhance biologically mediated purification processes.

The supply of drinking water to any major city is an enormous and complex task, but one that generally arises piecemeal as the city grows and merges with outlying suburbs. Providing a consistent supply of high-quality water to meet the needs of millions of households does not just happen by chance.

A dramatic illustration of the economic value of natural, ecosystem-mediated water storage and purification processes is the case of the water supply to New York. The city's Department of Environmental Protection delivers over 1.2 billion US gallons (4.5 billion litres) of water daily to 9 million people.

Early residents of New York drew water from wells and smaller water bodies, but the mushrooming population throughout the nineteenth century necessitated development of a network of aqueducts and reservoirs, bringing in water from sources of high natural quality a considerable distance from the city. In 1905, New York City looked northwards once again in its quest for more high-quality water, identifying the Catskills Mountains as a prime resource. Up through to 1928, various reservoirs and dams were constructed throughout the Catskills. Faced with ever-growing demand, the City had also turned in 1927 to sources in Delaware County and, after some legal wrangles, construction of the Delaware component of the Catskills–Delaware system (known as the Cat/Del system) was implemented in stages between 1937 and 1964.

The consequence today is that New York City has the largest unfiltered surface water supply in the world, delivered by natural capture and purification processes throughout a catchment of 2,000 square miles (830,000 hectares). The economic value of this is astronomical, when one considers how much it would cost to abstract, pump, store and purify water from the contaminated and enriched lowland rivers close to New York City.

Traditionally, the city had relied on sparsely populated rural catchments across the Catskills, where low-intensity use of land did not compromise raw water quality. Yet, by the 1980s, industrial-scale agriculture was replacing traditional methods and residential development added to the threats in these environmentally vulnerable areas. Contemporaneously, public health standards were becoming more stringent, specifying that all public water supply systems should provide filtration or else meet a higher level of water quality, operational and catchment management targets. In response, New York City implemented a comprehensive 'watershed protection programme' with mandatory standards for raw water, service to customers and source management. At the time, less than 30% of the city watershed area was in public ownership, increasing potential risks to the quality of raw water from the actions of intensive agriculture, industry, highways, residential areas, forestry and tourism activities.

Faced with the potential to implement expensive filtration plant – the 1990 estimated cost of a filtration plant was between $4 and 6 billion (£2.1 and 3.2 billion) plus annual running costs of over $200 million (£160 million) – the city's planners began to think in broader terms about protecting the resource, thereby averting massively escalating costs for water and sewerage services. Cost–benefit analysis suggested that a comprehensive programme of watershed protection would cost substantially less than filtration, but making this work was going to be a tough challenge.

Top-down, punitive regulations to force such systems to work had generally ended in failure elsewhere. It is for this reason that the city instead opted for a mutually beneficial urban–rural watershed protection partnership, providing simultaneous benefits to the residents of New York City and the communities in the Catskills and Delaware catchments who managed the water-yielding landscape that delivers the 'goods and services' enjoyed by their urban neighbours. Through a process of dialogue and consensus-building, farmers were educated about the environmental and economic risks associated with some farming methods while, in turn, the farmers educated the city about the economic pressures they faced and how previous anti-pollution measures were not workable.

By the end of 1991, the city and the farmers had begun to implement an urban–rural watershed protection partnership. This included the concept of the Whole Farm Plan that integrated agricultural pollution control into individual farm business plans. The Whole Farm Plan is a partnership also entailing input from a range of statutory natural resources, soil and water conservation bodies as well as agri-environment and other agencies and the farmers themselves. The public bodies carry out research into environmental risks. From this, they work in collaboration with farmers to develop an economically efficient plan for farming methods that minimise these risks. The city pays for staff inputs and the capital costs of pollution control investments on each farm as an incentive for farmers to join the programme.

Within 5 years, 93% of farmers in the Catskills had joined the Whole Farm Plan programme. Today, farm participation is still increasing, including 95% of farms, reducing agricultural pollution by 75% and stabilising the economics of farming in the Catskills. With the right mind-set, there is benefit for all to be gleaned from farming for clean water. Parallel partnership initiatives have since been undertaken between the city and forestry interests, land acquisition programmes and ecologically based land management. The process was completed in January 1997, when the constituent parties formalised all prior agreements within a comprehensive Memorandum of Agreement, to which the city committed funds of approximately $350 million (£190 million) in addition to the costs of various other initiatives in the watershed. The total cost of the watershed protection programme is approximately $1.3 billion (£700 million), which will maintain the city's pristine water quality for the foreseeable future.

A truly partnership-based approach was key to success, with involvement from and benefits for all key players, particularly for those who manage the land that delivers the ecosystem service of water capture. The linkages between rural and urban stakeholders has been important, as other watershed programmes elsewhere across the United States had served predominantly to alienate local residents and undermine the intentions and potential benefits of the programmes. The unique resource that supports the economic future for the city and wider region is the natural environment of the Catskills and the ecosystem services that occur there. Today, the dual goals of economic development and sustainable water quality are intimately intertwined.

Broadening the concept of farming from production of commodities alone into a model that recognises the value of water production, landscape and other ecosystem services is intellectually satisfying. However, the social and economic innovation to make it happen in practice requires advanced skills of vision, leadership and co-operative governance. The story of water supply to New York is one from which we should all take inspiration and the sense of great things being achievable when the ecosystem is placed at the heart of decision-making. It is, after all, the ecosystem that produces the goods and services that we use, and which forms an independent reference point for all parties potentially benefiting from enlightened institutional and economic relationships.

We are beginning to see some of this change in attitude to the outputs of land use in the 2002 revision of the European Union's Common Agricultural Policy (CAP). One of the key principles enshrined in the 2002 CAP revision was a shift of subsidies from production of commodities (so-called 'Pillar 1') onto agri-environment support ('Pillar 2'). The British government has rolled out implementation of the 2002 CAP reforms from 2005 under the new 'Environmental Stewardship' farm payment scheme. Instead of tying support payments to farm outputs, subsidies are now related to wider societal benefits arising from appropriate use and management of farmed land. These wider environment and landscape-scale benefits may include heritage, biodiversity and a variety of other facets of ecosystem functioning. As an already-developed society, our pace of change may be slow. However, we can all take heart and encouragement in seeing how a massive trading bloc such as Europe can take practical and significant steps in the positive direction of biodiversity- and ecosystem-centred thinking.

There are exemplars of protection of biodiversity and ecosystem functioning closer to home in the UK. When our urban water supply services were being implemented in the late Victorian period, a number of great metropolitan authorities had the foresight to buy tracts of largely agriculturally unproductive uplands as water capture, storage and purification zones. Although the UK's water industry has changed form over the century, becoming privatised in 1989, the legacy of this foresight remains. Great tracts of Wales, the Pennines, Cumbria and other parts of the country remain in the ownership of the water industry for the primary purpose of source protection.

As is the case in New York and other great cities worldwide, for example Sydney, Melbourne, Adelaide and Perth in Australia, changes in land use, industry, population and climate each pose a threat for the production of water of adequate quality and quantity from source protection zones. In many cases, agricultural activities undertaken by tenant farmers have jeopardised the quality of water abstracted by the very water service companies from whom they lease the land. When this occurs, the economic and social costs of loss of a water abstraction source vastly outstrip revenues to the water company from farm rentals, and can compromise security of supply to those served by the water utility. The mass of the population downstream and the water service providers that serve them are the net losers of degraded biodiversity and ecosystem functioning, while individual farmers alone reap relatively minor benefit from agricultural returns from unproductive uplands. Society, ecology and the wider economy are poorly served by such minimal financial gains accruing to a few yet compromising value and long-term security to the many. There is a range of published case studies exploring the respective costs and benefits of revised, more ecologically benign farm management. A particular small farm in Cumbria (north west England) shows nearly an order of magnitude greater societal benefits from a more sustainable form of land management [1].

It is in this new era of environmental and resource threat and opportunity that the multi-utility company United Utilities, the major provider of water and sewerage services in the north-west of England, set up SCaMP [3], the Sustainable Catchment Management Programme, instigated in association with the Royal Society for the Protection of Birds (RSPB) from around 2002. United Utilities owns 57,000 hectares (140,850 acres) of land in the north-west of England, which it holds principally to protect the quality of water entering reservoirs and rivers. Much of this land supports nationally significant habitats for animals and plants, with around 30% of the total area designated as Sites of Special Scientific Interest (SSSI) for its biodiversity value.

SCaMP is innovative in many ways. It builds on United Utilities' active catchment management programme, which recognises the role of biodiversity in producing high-quality water. However, it takes this several steps further. The first round of SCaMP, which started formally in 2005, targets two key upland areas in north-west England: Bowland and the Peak District. First, SCaMP places ecological processes at its very core, recognising that fully functional upland catchments under the ownership of United Utilities are the core resource for production of high-quality and reliable flows of water for ecosystem maintenance, abstraction and dilution of treated effluent downstream. Second, SCaMP identifies that habitat restoration can simultaneously benefit scarce ecosystems, for example the rare hen harrier (*Circus cyaneus*), the main English breeding site for which is on these United Utilities landholdings. This biodiversity is not only a co-beneficiary of restoration but also part of the functional ecosystems that 'produce' the clean water. Third, by altering land management agreements with tenant farmers, water production can be compatible with selected methods of agricultural production. Fourth, the reinvestment of water service charges into upland restoration is cost-effective in terms of averting the costs of increased levels of treatment, as well as shortages, lower in the catchment.

SCaMP aims to develop an integrated approach to catchment management within which to deliver government targets for SSSIs and wider enhancement of biodiversity, ensure a sustainable future for the company's agricultural tenants, and protect and improve water quality. Getting the proposal for SCaMP accepted by government and the environmental and financial regulators of the British water industry was an ambitious and strongly fought challenge, for which the United Utilities team can take considerable credit for their insight, scientific backup, development of strategic partnerships and sheer dogged determination. SCaMP puts ecologically based concepts at the heart of the application for investment, underlining corporate commitment to a sustainable form of water resource development. This is exactly the kind of advice more enlightened experts would generally advocate in developing countries, but which has often been lacking at home. The key difficulty was selling this progressive, uncertain and less familiar approach to government and its regulators through the periodic Asset Management Planning (AMP) process, which sets water service bills and which allocates the proportion of this revenue that may be invested in infrastructure and environmental schemes. In 2004, the water industry's financial regulator OFWAT, the government department Defra (the Department for Environment, Food and Rural Affairs) and the environmental regulator the Environment Agency finally agreed the AMP terms under which United Utilities could fund the SCaMP programme in the two target catchment areas. (United Utilities' original proposal had also included two additional areas in the north-west of England.) United Utilities had in any event commenced this work voluntarily since earlier in 2004 [2].

In the two accepted target areas in Bowland and the Peak District, funding was set aside to fence off vulnerable watercourses, move lambing off the moorland and develop Integrated Farm Management Plans for more environmentally sympathetic land management. United Utilities staff also worked with tenants to co-develop and agree plans, negotiate tenancy agreements and secure necessary agri-environment grants. There was also funding available to invest in any necessary one-off infrastructure improvements, as well as targeted planting of broad-leafed trees and selective monitoring of biodiversity and water quality. The RSPB was a partner in drawing up management plans sympathetic to wildlife and water production. A significant element of blanket bog restoration on the moorland top includes projects to block moorland 'grips' (drainage channels cut into the peat) that were draining the moorland with loss of characteristic habitat and oxidation of humic matter, which creates coloration problems in water abstracted downstream. By taking out this coloration at source, through rewetting of the moorland and moss regeneration, blanket bog habitat is being restored with all of its attendant ecological benefits while simultaneously delivering, it is to be hoped, higher-quality water and more reliable flows.

That, of course, is the theory. The timescale over which SCaMP will yield results is uncertain but is unlikely to be immediate. Nevertheless, early results are encouraging. Time will tell the extent and duration of the water quality, hydrological, ecological, habitat, fishery recruitment, landscape and associated tourism and other benefits that will accrue from a water service customer (i.e. public) reinvestment of approximately £10 million spread over 5 years, supplemented by relevant agri-environment subsidies. This significant investment in the biodiversity and ecological processes should eventually not only supply clean water but also help protect and enhance the characteristic landscape of the region, which is a significant 'draw' for the regional tourism industry. It is too early yet to say what the long-term outcomes for wildlife and water outputs will be, though early indications are that vegetation quality is already improving.

In the interim, SCaMP is an exemplar of a 'win–win–win' scenario for biodiversity, society and economic performance. The water industry depends on the goods and services of primary ecosystem functioning: clean water and the dilution potential of aquatic systems. Therefore, restoration of the aquatic environment has advantages for the ecosystems that 'produce' those resources. Biodiversity is set to benefit directly, including hen harriers, grouse, trout in upland streams, etc., while the demands of public supply are met on a more sustainable basis and should also include more dependable flows and reduced costs, energy and chemical inputs for treatment downstream. United Utilities is bullish about the scheme and hopes to extend the approach to further of its landholdings when funding can be released. Speaking of the programme to date in *SCaMPnews* [3] Issue 2 (the Spring 2007 edition of the newsletter of the SCaMP programme), Clive Elphick, Group Policy Director noted:

SCaMP is one of United Utilities most innovative programmes. It involves cost effective treatment of drinking water at source and is helping to reduce our greenhouse gas emissions. I hope we will see a second SCaMP programme for the period 2010–2015.

The relationship between society, biodiversity and business is closely tied and mutually beneficial in this instance. By restoring habitat for biodiversity, the uplands of north-west England also benefit from improved hydrology, water quality, landscape and amenity, retaining the character, resilience and economic viability of traditional landscapes.

Wessex Water [4] is another British water company that has allied itself to conservation and restoration of the water environment that 'produces' the goods and services on which it relies (clean water and capacity to reassimilate treated wastewater). One of the corporate values and public commitments of Wessex Water, the private supplier of services including water and sewerage to over 2.5 million customers within the south-west of England, has been to achieve sustainability. It was one of the first water service companies to make this commitment, was among the first companies to issue a sustainability report, and was the first to base its biodiversity actions on the UK's Biodiversity Action Plan (UK BAP). To further its commitment to deliver for biodiversity, Wessex Water has implemented partnerships with various wildlife organisations, including the Wildlife Trust, RSPB and Plantlife, and funded various biodiversity programmes on its landholdings and across the wider region that it serves. It has also sought proactively to reduce the impacts of its operations on nature. The commitment to finding a responsible relationship with biodiversity comes not from a public relations exercise, but from deep-seated corporate values and intentions as evidenced by practical action to deliver the aspiration. Also, of course, the 'enlightened self-interest' of healthier catchment ecosystems provides the company with more 'headroom' within one of the business's core resources. This pursuit of enlightened self-interest from more sustainable behaviour is a feature of a successful company understanding its interdependency with the ecosystems that support it and the people it serves.

Both the 'clean' and 'dirty' sides of the water business depend on the natural resource of clean water for abstraction and dilution. It is in the interests of United Utilities, Wessex Water, New York City and indeed other water service providers and users for biodiversity to 'work' and catchments to function efficiently, delivering clean water with a dependable flow. It is also a bold step that places ecosystem functioning at the core of business strategy for both sustained profit and public water supply.

Some other national initiatives place ecosystem services centrally in land management decisions. In as arid a land as South Africa, where water is predicted to be the single biggest future development constraint, initiatives that maximise the ecosystem service of water production from landscapes have significant social and economic value. The cost-effective delivery of water savings within South African catchments has been proven by the Working for Water programme initiated in 1996 and based on clearance of water-hungry invasive vegetation (see DWAF [5] and Woodworth [6]). Although funded from a job creation budget, Working for Water has significantly increased water yield from target catchments where invasive, water-hungry vegetation had compromised water availability for beneficial human uses. Preliminary assessments of the costs, benefits and progress of Working for Water demonstrate a considerable set of benefits associated with improved water yields resulting from clearance of invasive vegetation. The demonstrable success of Working for Water is seen as influential in the decision by former US President Bill Clinton to initiate the Comprehensive Everglades Restoration Program, one of the largest natural capital restoration projects in the world [7]. Related initiatives such as the Australian Landcare Scheme [8] and local projects set up by the UK's network of voluntary River Trusts (as reviewed by Everard [9]) demonstrate their effectiveness by putting the functioning of catchment ecosystems at the centre of planning to improve hydrology, water quality and beneficial services enjoyed by catchment communities. These examples highlight the importance of an ecosystem-centred basis for planning and, conversely, the likely range and scale of damages to societal interest when aquatic ecosystems are degraded.

The above measures address the protection or enhancement of the underpinning biodiversity that 'produces' ecosystem services constituting an essential primary resource essential for cities and businesses. However, they largely do so without creating a market linking the 'producers' and 'consumers' of these services. There are, however, emerging initiatives around the world that implement a Paying for Ecosystem Services (PES) approach founded on creation of markets. This is most advanced in South Africa, where innovative water laws, instituted as the country emerged from apartheid into democracy during the late 1990s, enshrine the principles of equity, sustainability and efficiency. This has promoted a range of pioneering approaches to adaptive governance, accounting for implications for the range of ecosystem services in development decisions, and detailed economic studies of the market value delivered by catchments together with the marginal implications of different development options. (For details of a practical spatial planning decision support tool for the city of Durban based on the environmental capacity of its

tributary rivers see Diederichs *et al.* [10] and for more on a study of economic impacts on ecosystem services in the Thukela (Tugela) river catchment see Mander [11].) A subsequent study in 2007 by the Maloti Drakensberg Transfrontier Project [12] seeks to make linkages between the restoration and management of upper catchment areas for the purposes of increasing run-off of water (building on Work for Water and related Working for Wetlands, Working on Fire and Working for Forests initiates), yielding economic benefits to communities lower down catchments and proposing a market mechanism by which heavy water users (forestry, intensive agriculture particularly sugar production, mining, industries such as paper mills, etc.) can invest in work to increase the water yield of the catchments on which they depend. This market model is finding favour with the South African government as a market means to embed the ecosystems approach as a basis for the equitable, sustainable and efficient provision of water.

Other PES initiatives are finding favour around the world, for example in the Lake Naivasha basin in Kenya, implicitly in the intent of the UK's Environmental Stewardship and the EU CAP agri-environment regime to maximise public benefits from land management, and novel approaches to flood risk management that step back from merely defending drained land at any cost and instead seek an optimal balance of ecological, social and economic benefits in novel schemes. In terms of more accurate reflection of the value of ecosystem services within our market economy, PES offers great hope for a more sustainable relationship with biodiversity. It also offers participating businesses a basis for securing the core resources on which their stability and progress depend.

4.2.2   The wood from the trees

Forestry products, too, represent a suite of industries close to biodiversity. We have already seen the economic, flood-related and other disasters that can occur when forest ecosystems are degraded or destroyed. This can even be at the national scale, as in the case of Sweden in the early twentieth century. However, there are also some inspiring tales of success where forest-related businesses have been at the forefront of protecting the core biological resources on which long-term profitability and security ultimately depend.

In the 1980s, graphic images of large swathes of clear-cut or burning forest in Canada, the Pacific Northwest of the USA and the Amazon were featuring with increasing frequency on television sets around the world. Ninety per cent of global forests lie outside protected areas, and were being increasingly felled to generate predominantly foreign revenue. Momentum was growing for timber boycotts, although it rapidly became clear that boycotts were not working as, in the absence of revenues from timber and other forest products, developing countries were inclined instead to fell their forests for production of beef, arable crops or other agricultural commodities to bring in foreign revenue. It was this climate of concern that gave rise to the notion of labelling timber from forests managed according to specific standards. A number of schemes were initiated.

The Forest Stewardship Council (FSC) was set up in 1994 by a consortium of interests that, at the time, seemed highly improbable [13]. This grouping of interests ranged from environmental NGOs (WWF, Friends of the Earth, Greenpeace), indigenous forest dwellers, professional forestry interests and major retailers such as Sweden's IKEA and the UK's B&Q. All shared a desire for a workable system that would promote responsible forest management practices and create a clear market for them. This bold, ground-breaking leadership created a scheme that was not only well in advance of what national government or intergovernmental groups had achieved, but seemingly also what they could conceive. The business drive was in fact essential to make the FSC scheme workable.

The FSC operation is relatively simple (at least in theory if not in operational reality), independent and transparent. The name 'Forest Stewardship Council', the acronym FSC and the FSC logo are all registered trademarks, which may not be used without prior authorisation. Only certified products may carry the FSC label. Forest owners and managers apply for the independent FSC auditing process to certify that their forests are managed to ensure long-term timber supplies while protecting the environment and the lives of forest-dependent peoples. A 'chain of custody' system traces forest products from source through the supply chain and eventually to consumers.

Today, FSC accreditation is a marque that enables consumers to buy forest products of all kinds with confidence that they are not contributing to global forest destruction. This includes FSC-accredited timber products and processors,

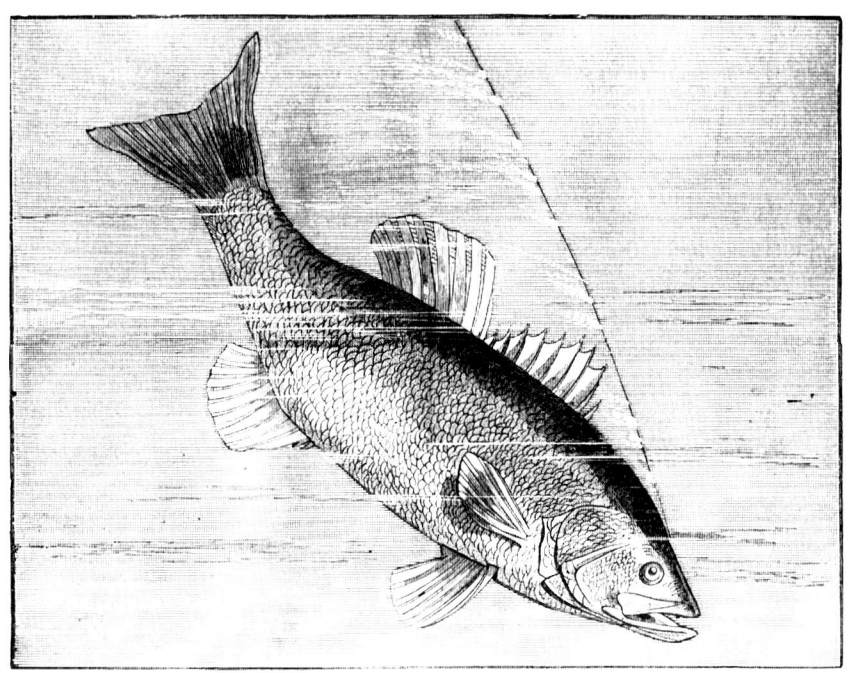

paper producers, printers and publishers, importers, retailers, architects, spec-ifiers, self-builders and more product streams besides. Thanks to the FSC, an increasing proportion of the wood used by industry and for consumer products originates in sustainably managed forests. As is clear from the many powerful interests who have bought into FSC, there is sufficient business advantage to sustainable forest management to justify the commitment.

As of November 2005, 67.16 million hectares of forest were FSC-accredited, indicating that they have undergone independent evaluation by one of several FSC-accredited certification bodies.

No system is perfect, of course. FSC has some critics who say it does not go far enough. However, after a decade and more of operation, FSC is a dramatic example of successful collaboration by diverse interests with a common purpose in sustainable forests and forestry industries. More needs to be done to evolve with changing knowledge and politics, the rigour of enforcement down supply chains, and the shifting use of forest-based products. However, the greatest chal-lenge remains to extend the market for certified products such that sustainability becomes the societal norm. The campaigning environmental NGO Greenpeace claims that 25 million acres of ancient forest are still being destroyed around the world every year.

4.2.3   Fishing for the future

Co-founded in 1996 by Unilever and the WWF, the Marine Stewardship Council (MSC) is a not-for-profit organisation modelled on the successes and processes of the FSC [14]. The benefits of sustainable fish stocks for Unilever, one of the world's largest manufacturers of food and household goods, are clear. It is esti-mated that Unilever purchases around 20% of Europe's fish catches.

The goal of the MSC is to achieve solutions to the problem of declining fish stocks, with sustainability as the key criterion. As discussed in the consideration of fisheries in Part III, the politically based quota system is often incompatible with sustainable management of fisheries, and the ecological and economic con-sequences of overriding fishery carrying capacity, often driven by subsidies, can be serious. The MSC seeks to promote fishery practices that are simultaneously environmentally and economically viable, and which maintain the biodiversity and ecological processes of the marine environment.

The MSC has been constituted as an independent, global, not-for-profit body since 1999. MSC accreditation is increasing the likelihood that fish consumed originates from a sustainable fishery, with a similar 'chain of custody' concept as used by the FSC to audit the transfer of fish from fishery to plate. Around the world, fish is a staple part of the diet for literally billions of people, and the fishing industry provides employment for an estimated 200 million people. Neither ben-efit will be sustained where fish stocks decline sharply or crash precipitously. It is therefore in the best interests of conservationists, fishermen, wholesalers, retail-ers, consumers, as well as industries utilising derivatives such as fish oil, fishmeal

BUCCANEERS CAPTURING A GALLEON.

and other marine-based products, to secure the sustainability of the ecosystems that support their livelihoods.

The MSC's strategy has been to develop and promote a generic global standard for fisheries, backed up by independent certification and product labelling. Research underpinning the accrediting of a fishery includes an assessment of the fish stock and the impacts of fishing, including the fishing methods used, on the target population as well as the wider ecosystem. The MSC may certify either individual fisheries or bodies such as associations of fishermen or producers. If the certification is granted, fish or fish products from that fishery may carry the MSC logo. At this point, the 'chain of custody' from fishery through to retailer comes into effect and, if approved, the product can carry the MSC logo.

Certification is a purely voluntary activity, but consumer pressure is considered to be creating impetus for more certification. Furthermore, MSC certification may avert the need for reactive regulation where fisheries become over-exploited. In any event, if the MSC marque contributes to the sustainability of marine fisheries, it is more than welcome. Some major players are backing the scheme in tangible ways. For example, Unilever, co-instigators of the MSC, has been not only progressively phasing out more problematic and threatened species such as North Atlantic cod from its ranges of seafood but also progressively phasing in MSC-accredited fish such as sustainably sourced pollock for its fish fingers from August 2007.

Like the FSC, the MSC is not perfect and has its critics. However, as an instrument that enables interested parties to participate in markets dealing in fish derived from sustainably harvested or farmed sources, the MSC creates a consortium of interest that encourages a sustainable relationship with biodiversity. Akin also to the FSC, the major deficiency of the MSC marque is lack of wider uptake; many more fish are still traded without regard to the integrity and sustainability of fish stocks and the ecosystems that support them.

4.2.4   Trading for tomorrow

Biodiversity in its many forms constitutes a directly tradable resource. It may provide a valuable source of foreign revenue particularly for developing countries. However, as for any other stock of species, over-exploitation can be disastrous for the traded organism, the ecosystems that sustain it and the human well-being that depends on sustainable trade. Annually, international wildlife trade is estimated to be worth billions of dollars and to include millions of plant and animal specimens. The wildlife trade is diverse, ranging from live animals and plants to a vast array of wildlife products derived from them. These include food products, exotic leather goods, wooden musical instruments, timber, tourist curios and medicines. Because the trade in wild animals and plants crosses borders between countries, the effort to regulate it requires international cooperation to safeguard certain species from over-exploitation.

As long ago as 1963, concerns were expressed at the United Nations about the impact of the international trade of flora, fauna and their parts or derivatives on long-term conservation of biodiversity. There was a clear need for binding international agreements to control over-exploitation of endangered species through international trade. It is from this international concern that CITES, the Convention on International Trade in Endangered Species of Wild Fauna and Flora, came to be agreed in 1973. As of 2006, there were 169 signatory nations to CITES [15].

From ivory to ebony, rhino horn to black coral, basking sharks to tigers, seahorses to mahogany and porcupines to Asian turtles, CITES has become an important instrument for the conservation of the world's species most endangered by human trade. CITES is the only widely recognised, respected and implemented international instrument that deals with sustainable international trade in wild species. Some species still remain heavily traded, vulnerable and not effectively managed, so the work of CITES on positive management and conservation has never been more urgent. Although CITES is legally binding for signatory parties, it does not take the place of national laws. Rather, it provides a framework to be respected by each signatory country that has to adopt its own domestic legislation to ensure implementation at the national level. Often this is effected at points where people and goods enter a country, particularly through the customs and excise infrastructure.

The basic text of CITES is supported by three appendices. The best known, Appendix I, prohibits any commercial trade in species that are already endangered, such as tigers, gorillas or the coelacanth. Appendix II includes about 95%

of the 30,000 species listed under CITES, and deals with species that are not yet endangered but may become so if trade is not regulated. A species appearing in Appendix II must be properly monitored and regulated to ensure that any trade is licensed within sustainable limits, and that specimens or their parts are derived from legal sources. Appendix III includes species for which international cooperation is required to prevent unsustainable or illegal exploitation in the plant or animal's home range, where domestic conservation legislation may apply.

CITES augments a range of national, supranational (such as European) and international legislation aimed at conservation of biodiversity, and is an example of cooperation over trade at the global scale to promote the integrity of biodiversity at local scale.

4.2.5   Eating well

The rise of organic agriculture is also related to business responding to public concern about the production of food in ways that do not contribute to declining biodiversity, intensive use of energy and synthetic chemicals, and residual contaminants in produce. Although organic food is the most well known and established, a range of other food-related initiatives – the Wholesome Food Association, local farmers' markets, Fair Trade and the like – embody some or all of the principles of farming methods sympathetic to biodiversity, averting excessive 'food miles', and fair recompense for producing communities.

The word 'organic' refers to biological origins, and therefore all food is technically 'organic'. However, over the past half-century, the term 'organic' has been used specifically to describe food grown without most types of artificial fertiliser or pesticide, emphasising crop rotation, using natural fertilisers, protecting refuge habitat for the natural predators of crop pests, and maintaining the life of the soil. The rise of organic agriculture was driven by consumer concerns and the network of agricultural, wholesale, retail and catering businesses serving it. Once again, business has driven a scheme with biodiversity concerns at its core, which has created new markets and opportunities.

Various codes applied by members of voluntary bodies established standards for organic produce. However, in 1993, a European Community Regulation became effective describing the inputs and practices that may be used in organic farming and growing, and the inspection system that must be put into place to ensure this. The Regulation also applies to processing and ingredients in organic foods. Since this time, all food sold as organic must come from growers, processors or importers who are registered and subject to regular inspection. The UK government (Defra, the Department for Environment, Food and Rural Affairs) developed the *Compendium of UK Organic Standards* [16], which described the procedures to be followed, and defines organic farming thus:

> Organic production systems are designed to produce optimum quantities of food of high nutritional quality by using management practices which aim

to avoid the use of agro-chemical inputs and which minimise damage to the environment and wildlife.

This principle covers:

- working with natural systems rather than seeking to dominate them;
- the encouragement of biological cycles involving microorganisms, soil flora and fauna, plants and animals;
- the maintenance of valuable existing landscape features and adequate habitats for the production of wildlife, with particular regard to endangered species;
- careful attention to animal welfare considerations;
- the avoidance of pollution and
- consideration for the wider social and ecological impact of the farming system.

Organic food and farming is promoted and certified in the UK by the Soil Association (and the associated Soil Association Certification Limited) [17].

Biodiversity, then, is writ large in the production of organic goods. Inspection during production is a key element in ensuring the pedigree of organic products, and it is illegal to sell any food as 'organic' or 'organically grown' unless it has been produced in full conformity with the EU Organic Regulation by registered producers. (For products derived from organic produce, manufacturers may use up to 5% of certain non-organic food ingredients and still label the product as organic. However, genetically modified ingredients and artificial food additives are never allowed in organic foods.) The EU Organic Regulation ensures conformity with the marque across the European Union, and a limited number of countries outside Europe are now currently recognised as having an equivalent system enabling their food to be freely imported and sold within the European Union.

Proponents of organic food make various claims about its superior taste and value for health, beauty and lifestyle. However, it is also in its demonstrable benefits for biodiversity through sympathetic production methods that the initiative attracts significant public support. Many also view organic goods as premium products. Again, it is a consortium of interested businesses that created this initiative protective of biodiversity. Furthermore, organic production has been a growth area ever since its inception, and the trend shows no signs of slowing.

Even without going to the level of organic standards, there are many things that landuse businesses and those that depend on and influence them can achieve. Simple measures such as separation of 'clean water' drainage from roofs and 'dirty water' drainage from yards can significantly lighten the load on biodiversity with appropriate diversion and low-technology treatment. Obviously, it can also reduce fees for the discharge of contaminated water. Pesticide containment and best practice, or ideally its use only for treatment and not as a prophylactic, also reduces pressure on wildlife. Natural pest control can also feature as part of more sensitive farming practices, rather than through chemically intensive treatment with all of its associated risks to wildlife and human health. Likewise, the timing

of certain maintenance and other works can be revised so as not to disturb critical life stages such as nesting birds, hibernating reptiles or mammals, or spawning and growth of vulnerable amphibians and fish. Many more simple, practical and profitable measures are available for those seeking to use land in greater sympathy with wildlife and natural processes. Such practices are intrinsic to a more caring corporate culture and are also, as we will see in the next chapter, likely to promote attention and loyalty among at least some customers, particularly key customers of niche business. Furthermore, these measures contribute to improving overall environmental performance and thereby attract lower regulatory scrutiny and charges and enhance corporate reputation and perception of taking risk management seriously. Even one step down the line, for businesses that are users rather than producers for farmed goods, these benefits can also accrue from working with strategic suppliers for a more sustainable relationship with productive land.

For all its perceived faults, including its less than broad-ranging implications for farmland biodiversity, the UK's 'Red Tractor' marque does at least institute a set of practices across an alliance of farmers, processors, retailers and distributors who work together to maintain and raise production standards. In 2006, some £5.3 billion of trade occurred under the 'Red Tractor' marque of the Assured Food Standards organisation [18]. Although it leaves scope for improvement, this and other devices do at least begin to institutionalise mainstream commitments to do more for a sustainable relationship with biodiversity.

4.2.6   The hole story

Mining is not an industry with a green aura. However, for the foreseeable future society depends on it and benefits greatly from mined products as diverse as minerals, aggregates, metals, nutrients and radioactive materials. The sustainability challenges are greater for some sectors of the mining industry than for others, but there are certainly companies who are leading the way in addressing the impacts of mining on society and biodiversity. As one mining company executive said to me, 'people complain that we mine where there are lots of SSSIs, yet it is our mining activities that have created the SSSIs in the first place!' Balance in all things, then. There are clear benefits to finding a sympathetic interrelationship with biodiversity in all mining operations.

In Chapter 2.4, we have already met the UK's Aggregates Levy, which aims to redress the balance of mining impacts. Nevertheless, the recycling of a small amount of money into local schemes, however worthy, is clearly not the be-all-and-end-all of corporate responsibility to biodiversity from mining and quarrying activities.

Many of the world's major mining enterprises – Alcoa, Anglo American, BHP Billiton, Freeport-McMoRan Copper & Gold Inc., the Mining Association of Canada, Noranda, Rio Tinto and WMC Resources Ltd – have collaborated in production of the IUCN (World Conservation Union) and ICMM (International Council on Mining and Metals) report *Integrating Mining and Biodiversity Conservation: Case Studies from around the World* [19]. This report does exactly what is stated in the title, outlining corporate experiences of internalising biodiversity conservation in various parts of the world. These initiatives cover informed and innovative management through to site rehabilitation and ecosystem reconstruction, and community engagement and collaborative decision-making. One of the recurrent themes of the report is the formation of new partnerships, for example with university departments, government departments and wildlife and social NGOs, to engage with initiatives such as addressing priority conservation targets, promoting environmental education in local schools, and so forth.

As mentioned above, quarry workings can often become wildlife-rich after use, and this has to be planned carefully during the lifetime of the workings. Responsible mining companies seek not merely to minimise and mitigate but, where possible, to enhance the biodiversity and public amenity value of land across the quarrying cycle. A practical example here is Omya UK Ltd [20], which mines and processes calcium carbonate (largely chalk) for the chemicals, food and other industries. A specific example is provided in the Environmental Statement attached to the company's December 2005 planning application to extend their Station Quarry site at Steeple Morden in Cambridgeshire. It is the company's stated intent to undertake approved restoration schemes with the objective of increased biodiversity, with restoration proceeding in phase with quarrying and taking place close behind the advancing quarry face. Detailed processes include stripping of high-fertility topsoil separately for subsequent use in site landscaping, with lower-fertility subsoil stripped together with overburden and employed for low-nutrient chalk grassland restoration. Plans for site restoration include several discrete but regionally appropriate habitat types. Some areas of restored land will be sculpted to create woodland with topsoil spread to promote tree growth, whereas most of the spent quarry area will be converted to calcareous grassland produced by spreading the lower-fertility calcareous overburden and seeding it with an appropriate proprietary or locally harvested seed mix. Calcareous grassland restoration was chosen above the alternative possibility of regeneration of arable farmland because, in the words of the Environmental Statement, 'The proposed quarry extension lies within a very extensive area of agricultural enterprise, where the wildlife resources of the area are impoverished. It would seem more appropriate to devote restoration efforts to creating more of a scarce commodity than of a widespread commodity.' Furthermore, grassland restoration recognises changing priorities in agricultural policy, away from production and towards wildlife improvement.

Some geological exposures of interest at Steeple Morden will also be conserved on the basis of English Nature's 2003 guideline *Geodiversity and the Minerals Industry: Conserving our Geological Heritage* [21], with managed public access, taking account of health and safety issues, to create an amenity. In addition to this, areas of soil compacted by intensive vehicle movement during quarrying are being used to allow the formation of temporary water bodies valuable to breeding amphibians and birds. Overall, the quarrying and restoration phases are planned to be hydrologically neutral, presenting no disruption of the natural percolation of precipitation into the groundwater. The December 2005 Environmental Statement for Station Quarry concludes: 'Taken overall, it is considered a better and more sustainable approach to devote resources to establishing additional areas of scarce chalk grassland, rather than a small amount of agricultural land of only average ALC grading.' (ALC is the UK's Agricultural Land Classification.) Sensitive quarrying with appropriate restoration can then prove to be not merely neutral, but net beneficial to biodiversity over the quarrying cycle relative to the arable land that the quarried site replaces.

This is far from an exhaustive consideration of mining and quarrying. However, it does illustrate that at the interface of business and biodiversity, even the 'ragged edge' that one normally associates with mining, innovative approaches that protect, restore and potentially even enhance ecosystems are possible when the problems and opportunities are acknowledged and addressed.

4.2.7   So what can my business do?

The lessons arising from the above examples of the role of business in successful management of biodiversity are manifold. These comprise a series of steps.

The first and most important step in all examples is to recognise that it is biodiversity and ecosystem functions that are the providers of tradable and useful goods and services. This should be a subject for discussion at high level in the company, as it challenges traditional assumptions about sources of business risk.

The second is to recognise that carrying capacity is finite and that mutually assured diminution of biodiversity and human benefits is the only likely outcome of over-exploitation.

The third step is then a recognition that management of ecosystems to maintain stocks of biodiversity on a sustainable basis is to mutual advantage. There is nothing wrong with the realisation that 'enlightened self-interest' may be achieved where resource protection secures a resource and boosts corporate reputation; in fact, it provides a positive and more durable incentive for its protection through sustainable exploitation. The suite of ecosystem services summarised by the Millennium Ecosystem Assessment, listed in Part I, is a good place to start this assessment of the business dependency on biodiversity and to explore changes in environmental impact stemming from different options and potential mitigation measures.

The fourth step recognises that consortia of common interests, often from diverse sectors of society including business and whole business sectors, environmental NGOs, indigenous or local people, land managers or fishery interests and regulatory bodies, may be the most effective means to achieve this commonly beneficial goal. For the higher risk business impacts on critical natural resources, it may also be possible to explore with others options for paying for ecosystem services (PES) schemes that secure the viability of and prolonged access to those resources.

The fifth step is to realise that there is a wealth of science available to apply in determining what does constitute a sustainable yield of biodiversity from the environments that support it. Opinion alone is an uncertain foundation and may not lead to a defensible 'story' or an advantageous outcome.

The sixth step is to recognise that collaboration and consensus are more effective than top-down compulsion, and that others also will derive benefit from collaboration to protect a resource.

The seventh stage is to accept, with others, voluntary or binding agreements to secure the long-term viability of the element of biodiversity that the common interest supports.

The eighth step (as demonstrated by SCaMP) is that the traditional role of a regulator telling industry to act in a more responsible manner with respect to resource conservation may not work, and that at times it is for industry to petition its regulators and government for permission to act in a more sustainable way towards the biodiversity resources. In other examples (such as CITES), national and international government may have to take a lead, though the enactment of agreement requires the collaboration of business.

The ninth step, as exemplified by the FSC and the MSC, is that protection of biodiversity may serve a valuable market differentiation role.

The tenth step entails developing a transparent method of reporting, such that the enterprise is seen to be accountable and trustworthy.

The eleventh step is recognition that publicity can be a force for good, offering both market differentiation and corporate trust to the business and enhancing public awareness of the value of conserving biodiversity.

Finally, the twelfth step is acceptance that protection of biodiversity through responsible business is quite simply 'the right thing to do', for its own sake but with many collateral benefits such as staff morale and retention, corporate image and shareholder confidence. Sound resource stewardship is also a factor strongly favoured by investors and corporate rating agencies.

If your business is close to biodiversity, analogous to the examples above, the benefits of protecting the biodiversity resources on which its and your mutual future depends will by now be largely self-evident. It is then a matter of joining an existing consortium of interest, or else establishing one through the twelve-step process outlined above. It all starts by recognising both the problem and the long-term benefits of 'enlightened self-interest'.

In the short term, there are, of course, economic risks where competitors continue to ignore the costs to biodiversity of their business activities. For example, they may retain dependence on the cheapest available pulp or wood derived from old-growth Indonesian or Amazonian forests to undercut your prices. However, the world is changing, including public values and investors' attitudes to risk. In the internet era, the risk of exposure of unethical or exploitative business practice is never more than a couple of mouse-clicks away. For the ethical producer, there is mileage to be had from the brand differentiation of a proactive strategy with respect to biodiversity. And, consequently, a little publicity of the issues of potential public concern may help reveal to customers the hidden costs of artificially cheap production 'subsidised' by the destruction of ecosystems and those dependent on them, and that it is only responsible to use suppliers who respect biodiversity.

Anyhow, surely it is a matter of basic risk management. It is certainly also 'the right thing to do', with all the assurance that this may offer of less exposure to resource scarcity, reputation, stakeholder pressure or staff attitude. Biodiversity is the lifeblood of business and society, and needs to be nurtured as such.

References

[1] Everard, M., Kenmir, W., Walters, C. & Holt, E., Upland hill farming for water, wildlife and food. *Freshwater Forum*, **21**, pp. 48–73, 2004.
[2] SCaMP, available at: www.unitedutilities.com/?OBH=3128.

[3] SCaMP, SCaMPnews 2, Warrington, UK: United Utilities, 2007.

[4] Wessex Water, available at: www.wessexwater.co.uk.

[5] DWAF, Working for Water Programme: Annual Report 1996/1997. Department for Water Affairs and Forestry, Pretoria, 1997.

[6] Woodworth, P.,Working for water in South Africa: Saving the World on a single budget? *World Policy Journal*, pp. 31–43, 2006.

[7] Everglades Restoration Program, available at: www.evergladesplan.org.

[8] Landcare Scheme, available at www.landcareaustralia.com.au.

[9] Everard, M., Investing in sustainable catchments. *Science of the Total Environment*, **324(1–3)**, pp. 1–24, 2004.

[10] Diederichs, N., Markewicz, T., Mander, M., Martens, A. & Zama Ngubane, S., eThekwini Catchments: A Strategic Tool for Management. First Draft. eThekwini Municipality, KZN, 2002.

[11] Mander, M., Thukela Water Project: Reserve Determination Module. Part 1. IFR Scenarios in the Thukela River Catchment: Economic impacts on ecosystem services. Institute of Natural Resources, Scottsville, 2003.

[12] Maloti Drakensberg Transfrontier Project, *Payment for Ecosystem Services: Developing an Ecosystem Services Trading Model for the Mnweni/Cathedral Peak and Eastern Cape Drakensberg Areas*, ed. Mander, M, INR Report IR281. Development Bank of Southern Africa, Department of Water Affairs and Forestry, Department of Environment Affairs and Tourism, Ezemvelo KZN Wildlife, South Africa, 2007.

[13] Forest Stewardship Council, available at: www.fsc-uk.info or www.fsc.org.

[14] Marine Stewardship Council, available at: www.msc.org.

[15] CITES. Convention on International Trade in Endangered Species of Wild Fauna and Flora. www.cites.org.

[16] Defra (UK Department for Environment, Food and Rural Affairs), The Compendium of UK Organic Standards (www.defra.gov.uk/farm/organic/standards/pdf/compendium.pdf).

[17] Soil Association, available at: www.soilassociation.org.

[18] 'Red Tractor'/Assured Food Standards, available at: www.redtractor.org.uk.

[19] IUCN (World Conservation Union)/ICMM (International Council on Mining and Metals), *Integrating Mining and Biodiversity Conservation: Case Studies from around the World*, available at: http://www.icmm.com/page/1155/integrating-mining-and-biodiversity-conservation-case-studies-from-around-the-world.

[20] Omya UK Ltd, Planning Application to Extend Station Quarry, Steeple Morden (Letter and associated Environmental Statement dated 29 December 2005 to the Mineral Planning Officer, Cambridgeshire County Council).

[21] English Nature, *Geodiversity and the Minerals Industry: Conserving our Geological Heritage*, English Nature: Peterborough, UK, 2003.

4.3   One step removed

Many of the activities of most businesses operate at one step removed from the interface between primary biological resources and profit. However, even at one step removed, business activities still have a dependency on biodiversity and will be compromised if the underpinning natural resources decline. Could your enterprise function without the ecosystem 'goods' of water and air or secondary products derived from biological resources such as timber or wood fibre in the form of paper, card or construction products? How about the myriad goods derived from productive soils? And what of resources from natural deposits such as gravel and sand, or the oil and coal that drives those power generation plants into which your computers and machinery are plugged? And, for both their direct and indirect emissions to water, land or air, what is the consequence of exceeding the limits of ecosystems to safely disperse and assimilate waste matter?

Although we may not commonly think of it this way, and when much of our inherited and deeply unsustainable economic drivers almost completely externalise it, this broader 'footprint' on the natural world bears the weight of our enterprise. We might buy these things in without the awareness that we are using biodiversity, but when and if the productive ecosystems on which we rely break down, or damage to them is exposed in the media, we can be sure that we will not be immune from resource scarcity and price rises or the perception and reputation issues that will inevitably follow.

4.3.1   Protecting ecosystems enjoyed by customers

For a wide range of products, customers require and therefore depend on the goods or services produced by biodiversity. Therefore, although the products themselves may not derive directly from raw biological resources, 'outdoor' and sporting goods manufacturers and countryside sports services are among businesses that already understand that they are in fact close to biodiversity and depend on it for trade. Without fish to angle for, diverse and beautiful landscapes to hike or camp in, cycle through or otherwise enjoy, game to hunt and scenery to paint or photograph, these supporting businesses are not viable.

It is for this reason that some fishing tackle companies ring-fence a proportion of profit for reinvestment into the protection of vulnerable rivers and other waters. The Hardy & Greys fishing tackle company [1], based in the north-east of England, operates such a bursary scheme, the three strands of the 2007 programme including support for project work by the Wild Trout Trust (an excellent charity dedicated to the conservation of wild trout in Britain and Ireland through protection and restoration of their habitats) and for the environmental campaigning work of both the Salmon and Trout Association and the North Atlantic Salmon Federation. Likewise, the Orvis [2] fishing tackle and outdoor clothing company has supported such initiatives as the not-for-profit Riverfly Partnership's fundraising calendar for 2007 [3], as well as work undertaken by the Wild Trout Trust [4]. Many others British fishing and fishery businesses and interests support organisations such as the Anglers' Conservation Association (ACA) [5], which is a not-for-profit organisation that uses the common law as a means to protect aquatic environments from pollution and other harm. In the USA, the not-for-profit organisation Trout Unlimited has been active in river con-servation and rehabilitation with the support of a wide membership that includes many businesses with associated interests. In a similar vein, many British shoot-ing interests support the conservation works of the not-for-profit British Associa-tion for Shooting and Conservation (BASC) [6] while, in the USA, waterfowling interests might support the not-for-profit Ducks Unlimited that has had major successes in wetland creation and protection. Some outdoor pursuits clothing manufacturers support wildlife conservation initiatives for similar reasons. For this type of company, promotion of biodiversity is not merely a 'good thing' but also a strategic corporate investment into a primary resource underpinning con-tinued trade.

Lifestyle companies, too, have been active in their financial support for pro-conservation angling organisations. For example, Classic Malts of Scotland co-sponsors the Wild Trout Trust. The UK's Salmon and Trout Association [7] has also previously enjoyed the sponsorship of whisky distilleries. The reason for this may partly be for the sponsoring business to identify its brands with concerned anglers who are members of organisations promoting environmental protection for the benefit of fish stocks, but also because protection of salmon stocks in rivers automatically ensures the high-quality water required for distillery purposes.

Some businesses have benefited from corporate values and product offers that appeal to the wildlife conservation ethos of their customers. For example, the Yeo Valley Organic diary product company, based in Somerset, takes the organic ethos throughout its operations and farming enterprises [8]. The company plants and maintains broad-leafed trees on its farms to create wildlife habitat, is experimenting with the cultivation of biofuels, and is maintaining other habitat including drystone walls and unploughed permanent set-aside strips around field edges. In early 2007, it initiated an offer for its customers to collect product packaging to claim one or two tree saplings of native broad-leafed trees for planting out in gardens, or to use the tokens to support a donation to the Woodland Trust (a British charity leading on the preservation and protection of native woodland heritage) [9]. This creates brand identity but also directly supports pro-biodiversity action on the back of sales.

Other food companies are seeking to do good for biodiversity on the back of profit. One such is the Green & Black's organic chocolate company [10], which runs various initiatives including a direct relationship with the Maya Gold Project in Belize. The Maya Gold Project started in 2003 under funding from the UK government's Department for International Development (DfID) and co-funding by Green & Black's. The project has the aim of turning the Toledo Cocoa Growers'

Association (TCGA) into a viable, self-sustaining organisation. Obviously, social and economic empowerment is a major goal of this project, but the organic ethos also recognises that this progress is only possible from the foundation of a sustainable relationship with biodiversity and ecosystem functioning. Stimulation of ethical and environmentally responsible supplies of the raw resource of cocoa beans clearly has value to Green & Black's in addition to the benefits of consumer identity with those ideals.

Even the economic giants, often pilloried for their heavy-handed relationships with suppliers leading to adverse consequences for the livelihoods of farm businesses and for biodiversity due to downward pressure on farm gate prices, are increasingly seeking to identify with 'greener' consumers, building brand loyalty. Even such economic giants as the retailer Tesco try to gain customer attention and loyalty through such nature-friendly messages as their flyer *Little Steps to Becoming Greener*, distributed through UK stores in the summer of 2007. Ultimately, businesses will be judged on their depth of practical action, an increasingly sophisticated and environmentally literate public seeking the real impacts on the natural world and no longer bamboozled by rhetoric alone.

This, of course, takes us into the territory of 'cause-related marketing', which has its strong champions and its fierce critics. On the one hand, cause-related marketing, tying a brand to a social or environmental campaign, is a proven method for brand visibility and differentiation as well as a generator of customer attention and loyalty. Critics see cause-related marketing as giving some companies access to sectors of society on the back of shallow promises and/or less than generous returns to the charity or cause. Today, all types of consumers are demanding greater accountability and responsibility from corporations; done right, cause-related marketing can generate wins for the business, the cause and the sense of contribution to it by the customer. With appropriate commitments and safeguards, it is possible for positive steps for biodiversity to ensue from businesses of all types through cause-related marketing, from which brand value and customer satisfaction also result.

4.3.2   Bound by natural limits

Limitation of natural resources, particularly where there is competition for them from a burgeoning population and a competitive multinational market, will increasingly constrain our freedom to operate in the future. The ways in which this ever-constricting reality will first manifest itself in the 'real world', the one we plan for when we make commercial decisions, remains unpredictable. However, we can be sure that it will appear in one guise or another.

The most obvious type of business impact from depleting biodiversity is via resource scarcity and spiralling costs. As a simple example, timber prices reflect scarcity of exploitable forest and the extra effort required to harvest less accessible sources. This is particularly so for exotic timbers that society still, irresponsibly, permits to be traded at volumes well beyond their natural replenishment rates and the capacity of some types of forest to recover. Water bills too reflect

more stringent environmental controls on effluent quality for waste, as well as measures necessary to address 'upstream' water scarcity driven by rising societal demands on a finite resource diminished by both catchment functioning and climate change. Also, energy bills reflect increasing concern for environmental considerations as well as an effective cross-subsidisation of less affluent communities who have a right to connection to energy. The roots of all supply chains are ultimately in the soil of biodiversity, its products and/or services.

Biodiversity is part of the diverse tapestry of issues comprising the 'licence to operate' that society grants to businesses. Consequently, there are many other more subtle ways in which biodiversity impacts may eventually affect business.

These include, as just a few additional examples, corporate reputation issues, slow or tricky planning applications, scrutiny by would-be customers, the concerns of shareholders and corporate ratings agencies, and fiscal measures. Natural limits impose themselves on business operations every day, many through the route of taxation, though some also through subsidies and other market-based incentives. I'll explore a few recent tax changes in the UK to observe the sometimes hidden role of societal concern for biodiversity.

4.3.3   A changing climate for business

The 'greenhouse effect' is a natural phenomenon on this planet caused by the trapping of infrared radiation by certain atmospheric gases, which means that this planet is warmer than might otherwise be expected from its distance from the sun. This provides the conditions in which biodiversity can arise and prosper. However, global society's dependence on fossil fuels and other problematic chemicals produces gaseous emissions, which accumulate in the atmosphere and trap an accordingly greater amount of infrared radiation, like a greenhouse trapping the sun's heat. These amplify the natural variability of climate to a state where weather patterns, natural cycles, biodiversity and the climate from global to local scales are disrupted. Fossil evidence (such as gas and water locked away permanently as ice in polar regions) indicates that global warming was increased as society began to exploit fossil fuels since the beginning of the Industrial Revolution. It has accelerated ever since.

Throughout the twentieth century, the global climate warmed by approximately 0.6 °C, a mean of 0.06 °C per decade. Computer models now predict a range of possible climate futures with a median estimate projection of a further rise of 3 °C by the end of the twenty-first century, or 0.3 °C per decade. It is currently believed that most ecosystems can withstand at most a 0.1 °C global temperature change per decade; much beyond this and severe stresses and species extinction are inevitable. Predicted climate change scenarios would certainly shift current climate zones, allow the invasion of species outside their native ranges, and add to current habitat fragmentation with substantial net damage to ecosystems evolved to cope with or adapt to a more sedate natural pace of change in climate. The irony is that the greenhouse effect is an entirely natural process but, where humanity's actions override nature's capacity to absorb waste gases safely, it is nature that suffers first and in turn poses a massive threat to humanity.

Current pressures on business to pay attention to carbon management are therefore welcome. However, are they merely a political 'flash in the pan' that was neither foreseeable nor will be of any great longevity? Actually, no. Climate scientists and environmental pressure groups have been warning of the perils of climate change now for at least two decades. Early evidence was equivocal and the scientific theory was rudimentary, so the business implications both for companies and for politicians overwhelmingly concerned about competitive strength were easy to dismiss.

However, the issue of climate change did not just go away. Indeed, it did the exact reverse as the scientific canon of both theory and evidence grew to a solid consensus now denied only by a few 'experts' who are, in the main, sponsored by businesses with a vested interest in denial. Society-wide certainty about the likely human-accelerated causes of global warming has all but solidified. Despite some of the remaining climate change nay-sayers being close to political power, particularly in the United States, climate change had been rising up the political agenda throughout the world over a period of years. Climate change finally broke out in 2005 as a major international issue when people including political leaders realised the massive threat that it poses to our collective and localised wellbeing. This rising recognition and concern has since hardened into policy action, some of which we will look at later in this chapter. Public opinion has also hardened about the need to tackle our habit of overloading the global ecosystem's ability to reassimilate greenhouse gases, with 'blockbuster' movies such as *The Day After Tomorrow* and *An Inconvenient Truth* confronting it on a popular level. Businesses have been responding for some time, and we are seeing the dawn of some high-profile and goal-oriented corporate commitments such as Rupert Murdoch's News Corp. announcing in May 2007 a commitment to go 'carbon-neutral' by 2010. Business drivers for carbon neutrality show signs only of intensifying. Carbon management in the round is on the agenda for business, particularly in the UK. And, although climate change is a secondary effect of using natural resources imprudently, at its core lie the ecological effects of overloading the global biosphere's capacity to reabsorb and process waste gases.

But business is not necessarily an inactive passenger in this process, nor a dead weight resisting change as it is sometimes portrayed. The gigantic global insurance and reinsurance industry has had a major role in driving the climate change agenda. Indeed, it has had a busy few decades acting on and influencing the wide recognition that climate change is consequent from greenhouse gas emissions, particularly carbon dioxide emitted from combustion of fossil petrochemicals, gas and coal, and that one of the impacts is more severe and unpredictable storm and flood damage. Risks associated with ecological functioning, the action and response of biodiversity, lie at the heart of the business of insurance and reinsurance. Weather-related disasters cost US industry $70 billion during 2003, which accounts for the role of this sector of business as one of the main drivers of international pressure to address climate change for business as well as public benefit. Governments around the world are responding to the threat of climate change, each at a different pace, with the UK perceived as a leader. This is so now even in the United States, generally perceived as a bastion of entrenched self-interest and climate change denial, particularly since the ironic and substantial immobilisation of American petrochemical processing capacity by Hurricane Katrina and a suite of lesser storms in 2005.

A great deal of the credit for international recognition, research and ongoing consensus-building and responses to the threat of climate change can be ascribed to international collaboration. In particular, the continuing work of the Intergovernmental Panel on Climate Change (IPCC) has led international thinking and recommendations for action [11, 12]. The IPCC was established in 1988 by the World Meteorological Organisation (WMO) and the United Nations Environment

Programme (UNEP), comprising a number of the world's leading scientists in the field of climate change. The role of the IPCC is to assess and review scientific, technical and socio-economic information associated with human-induced climate change. Since that time, the IPCC has continued to develop and provide scientific, technical and socio-economic advice to the world community on the state of knowledge of causes of climate change, its potential impacts and options for response strategies. The authoritative IPPC report *Summary for Policymakers of the Fourth Scientific Assessment of the Physical Science of Climate Change*, published in February 2007, has all but ended any substantive debate about the evidence for climate change and the likely major contribution of human activities.

Global action to tackle climate change is as yet fragmented and disproportionate to the threat. The 1992 UN treaty on climate change was the first concerted international agreement, setting the target of a total cut of at least 5% in greenhouse gas emissions from 1990 levels by 2012. The Kyoto Protocol, named after the Japanese city where it was agreed at international conference in 1997, set greenhouse gas emission targets on a national basis founded on the 'convergence and contraction' principle. This sets reduction targets for developed nations such as the world's biggest emitter, the United States, while allowing some headroom for growth in emissions for developing countries that were on a pathway of industrialisation. Overall, the total global emission would decline, particularly so when industrialising nations then began to apply novel technologies to reduce their emissions again. The Kyoto Protocol finally came into force on 16 February 2005, its implementation delayed because of a requirement that countries accounting for 55% of the world's emissions had ratified it. This goal was only reached after Russia signed up to the deal in 2004.

To date, 140 nations have ratified the Kyoto Protocol, 30 of them industrialised nations. Nevertheless, the lax timetable indicates massive foot-dragging when it comes to practical action to back up an issue about which political leaders from most nations speak in the rhetoric of urgency. Of particular and deep concern is the continued absence from the deal of the world's largest emitter of greenhouse gases, the United States, which accounts for almost one-quarter of global emissions, as well as some other key countries such as India, China and Australia. In fact, the United States had initially agreed to a 7% emissions reduction before incoming President Mr George W. Bush denounced the pact in 2001 and led a move not to ratify Kyoto agreements to which the United States had signed up. Bush also famously refused at the Gleneagles G8 summit in 2005 to countenance any measures that might undermine the primacy of the US economy, and is also widely reported to have stalled international action to progress to further targets beyond the 2012 lifetime of the Kyoto agreements. Such is the legacy for the future from Bush and his petrochemically funded regime, with which biodiversity and humanity may have to cope. The denial and self-interest of the world's biggest emitter and other emerging nations sets a poor example of statesmanship and foresightedness, and can hardly be expected to bolster real commitment in other nations to tackle one of history's greatest self-inflicted threats on humanity's future. Our prior analogy of the responsible business being undercut by unethical practices comes to mind.

But, despite the fact that the Confederation of British Industry (CBI) in the UK makes all manner of dire predictions about harm to business if it has to respond to the major threat of climate change – apparently it is better to be competitive now and live in a world with no future or market for products! – all is far from doom and gloom for business. Indeed, some elements of business are again taking some of the pioneering steps.

The UK's *Guardian* newspaper on 15 May 2004 reported an institutional investors' survey of corporate attitudes, which found that the profile of climate change across businesses was growing strongly, even in the petrochemical giant ExxonMobil that had hardly been at the forefront of accepting the phenomenon or addressing it proactively. The majority of the 95 participating major companies in this survey said that climate change presented risks but also opportunities to their businesses. The *Guardian* report also observed that the world's most powerful investors have an obvious reason for wanting to avert climate change which would devastate their wealth. This of course preceded the consequences of Hurricane Katrina, and the major impact of visible devastation of the city of New Orleans on the American psyche. However, even before that, most major US business leaders were beginning to take climate change seriously. Spiralling oil prices have since further focused the minds of business on the wisdom of carbon management, particularly with the OECD (Organisation for Economic Cooperation and Development, the think-tank of the world's richest nations) favouring a shift to a low-carbon economy, meaning that the 'cost of carbon' is becoming a pressing matter for energy-intensive global businesses.

Also in the United States, the Pew Center's Business Environmental Leadership Council (BELC) was formed under the belief that business engagement is critical for developing efficient, effective solutions to climate change [13]. The Pew Center's contention was that companies taking early action on climate strategies and policy could achieve sustained competitive advantage over their rivals. The BELC is now the largest US-based association of businesses focused on addressing the challenges posed by climate change, with 40 members representing $2 trillion in market capitalisation and over 3 million employees. This membership covers many different business sectors from manufacturing, transport, oil and gas, utilities and chemicals. This diverse and influential US membership, despite America's reputation to the contrary under the Bush administration, recognises that the risks and complexities of climate change are so important that collective action is necessary to meet this challenge. All accept that enough is known in scientific terms to take action, and that businesses can and should take definite steps now to determine opportunities for emission reductions and new, more efficient products, practices and technologies. The members of the council are also committed to implementing the market-based mechanisms that were adopted in principle in Kyoto, and to encourage other territories to join in with the process. For the business leaders within the BELC, addressing climate change while sustaining economic growth in the United States are not seen as incompatible goals.

This view is consistent with that of the OECD which considers that, despite some short-term winners and losers across various business sectors, standing up to the opportunity of a low-carbon economy will be, at worst, economically neutral overall. Indeed, the OECD believes, it can be expected to generate all manner of new technologies, employment prospects and business opportunities. Subsequently, the UK's *Stern Review on the Economics of Climate Change*, published in October 2006, offered the most comprehensive review to date on the economics of climate change, with Sir Nicholas Stern commenting at the report's launch that 'Strong, deliberate policy choices by governments are essential to motivate change... But the task is urgent. Delaying action, even by a decade or two, will take us into dangerous territory. We must not let this window of opportunity close' [14].

So business is again in the vanguard. Business also needs to respond to a changing political and fiscal climate that brings climate change into the core of business thinking. One such practical UK government response includes the Climate Change Levy, introduced in the 1999 budget and coming into effect in April 2001, on business use of energy to encourage innovation to reduce energy demand. This is one of many measures that will be necessary to deliver UK government resolutions to tackling climate change, including a renewed commitment in March 2007 to 60% greenhouse reductions by 2050. This goes significantly beyond a prior promise of 20% reductions by 2020 and the UK's Kyoto Protocol target of a 12.5% reduction (relative to 1990 levels) by 2010. The EU Emissions Trading Scheme (ETS) is one of the policies introduced at European scale, effective from 2005, that places costs on business for CO_2 emissions but also creates the opportunity for profit from sale of excess credits.

Concern for climate change can also be big business. This is in terms of not merely the opportunities for consultants to work with big enterprises, but also the multi-billion-dollar emerging emissions trading industry. A report by the World Bank launched on 10 May 2006 records that the total volume of carbon traded worldwide in 2005 (696 million tonnes of CO_2) was nearly six times higher than in 2004 (119 million tonnes of CO_2), with the launch of the report coinciding with the opening of the world's largest carbon fair in Cologne, Germany. The launch of the ETS across the EU adds to obligations under the Kyoto Protocol in focusing the minds of business on the benefits of carbon management, with the lingering threat of government action to introduce increasingly stringent compliance instruments to address greenhouse gas emission reduction. The ETS has seen dramatic growth of nearly 3,600% in its start-up phase, rising from 9 million tonnes of CO_2 in 2004 to 322 million tonnes in 2005 representing an economic value of US $8.2 billion (€6.4 billion). Both the ETS and wider international trading mechanisms under the Kyoto Protocol are enjoying continued expansion, with the upward trend seemingly set to continue as climate change continues its transition from a marginal ethical concern to a mainstream business issue.

The Renewables Obligation in the UK (and its associated Renewables [Scotland] Obligation) provides further incentive for businesses to control their impacts on biodiversity through contributions to climate change. It is a UK government initiative that came into force in April 2002 as part of the Utilities Act (2000) to encourage a transition to climate-neutral power generation. The Obligation requires power suppliers to derive a specified proportion of the electricity they supply from renewable sources, starting at 3% in 2003 and rising gradually to 10% by 2010. Eligible renewable generators receive Renewables Obligation Certificates (ROCs) for each megawatt-hour (MWh) of renewable electricity generated, and these may be traded with other suppliers to enable them to fulfil their own obligations. ROCs have traded as high as £47/MWh and, although there is no guarantee they will remain this high, rising oil costs and a guarantee in law of the ROCs scheme through to 2027 suggest that they may be of high value. Certainly, for enterprises such as sewage treatment and chemical companies, generation on-site from methane combustion, combined heat and power plants and so on creates a valuable market for ROCs, which represents a stable revenue stream from measures sympathetic to biodiversity.

If there is a simple 'take home' message from the preceding discussion and examples, it echoes that in the Stern Review: the cost of inaction is likely to massively exceed that of proactivity.

4.3.4   Taxing times

Biodiversity and wider environmental concerns have also driven other shifts in fiscal policy. The UK enjoys the Aggregates Levy, described in Part I and, since 1996, the Landfill Tax, reflecting some of the cost of the habitat 'take' of landfill sites and the wider environmental repercussions of buried waste. It operates at the opposite end of the life cycle of materials to the Aggregates Levy. The Landfill Tax imposes a landfill site gate fee, with two differential rates for inactive and all other waste, both of which escalate beyond the rate of inflation. The Landfill Tax is an economic instrument creating a positive disincentive to landfilling of waste. This tax is distributed back into research and practical measures intended to boost the efficiency of resource use and the decline in volumes of landfilled waste. It is designed to encourage businesses to produce less waste and to use alternative forms of waste management.

Waste flows are also subject to a raft of EU Directives, which progressively cap the volume of putrescible and other organic waste entering landfill sites, the segregation and reduction of certain types of hazardous waste streams, and mandatory recycling targets. Financial penalties are applied to local councils failing to meet their contribution to national waste targets, and these costs in turn are handed on to businesses along with reactive measures to discourage waste arisings. Arguably, the public and environmental costs of illegal fly-tipping, significantly on the rise as waste legislation imposes greater costs and restrictions, are another charge borne by society for the waste it produces.

The reason for detailing these fiscal and other economic measures is simply to demonstrate that, although each highlights an impact on business from an activity one step removed from direct dependence on a biological resource, the pressures fed back to that business are no less direct or real. And, yet, how commonly do we consider these aspects of indirect dependence on biodiversity in risk management processes?

4.3.5   Treading more lightly

So the prior observations of self-benefit arising from leaving a smaller, shallower 'footprint' on the natural world are as germane to business activities one step removed from biodiversity, whether or not the biodiversity is visibly 'traded' through an enterprise or is used in the form of secondary products and services.

The implications for a review of procurement should already be clear. For example, what use could the business make of trusted sources such as FSC, MSC- or organic-accredited products (and other such credible marques) to replace existing paper, wood, forestry products, food and other resources currently bought without further scrutiny? Or how can the purchasing power of the organisation be brought to bear to ensure a greater scrutiny by its key suppliers of the pedigree of products and services procured, i.e. further up the supply chain.

And then there is a tighter focus on the measures that can generate quick payback, such as those that improve the efficiency of use of water, energy, aggregates and the reduction of waste flows of all types, which will bring immediate cost reductions and tax benefits allied to a lightening of the burden on biodiversity. This then places the well-known mantra of 'reduce–re-use–recycle' into the context of a responsible corporate attitude towards biodiversity. (I'll note here in passing that great attention is often paid to recycling and less on the first step – 'reduce' – from which maximal biodiversity and business benefit can arise.)

While a wider range of case studies is available today, most of them covering far more recent initiatives, an example of the value of resource stewardship from IKEA in the 1990s remains as simple and compelling today as it is generally applicable. A year before rethinking its approach to waste, the annual costs for waste management at IKEA's Gothenburg (Sweden) store was 250,000 Swedish Crowns (SEK) ($US 30,000). Then, after working with sustainable development charity The Natural Step, IKEA introduced the 'Trash Is Cash' programme, taking account of items that could be sold for scrap or phased out and also of the costs of the sorting process. The following year, the waste management bill for the Gothenburg store was down to 0 SEK. The next year, the 'waste stream' returned a profit of 40,000 SEK ($US 5,000). Profit, it appears, is to be picked up like litter if one focuses on the end-game of sustainability rather than waits, along with all of one's competitors, to be driven to act. IKEA is far from alone in reaping dramatic financial benefits from looking at the 'downstream' end of its business, with all of the demands and impacts on biodiversity that this implies.

The same principle of averting eventual waste from the downstream end, or in other words increasing the economic efficiency of the material throughput of the business, is also becoming apparent in the increasing legislative focus on end-of-life products. Pertinent examples include relatively recent EU Directives on the take-back of waste electrical and electronic equipment (under the Waste Electrical and Electronic Equipment Directive), mandatory recycling targets for automotive parts (under the End-of-life Vehicles Directive) and packaging waste (under the Packaging Directive). Planning upstream to avert wastage and consequent emissions downstream will have growing financial value to the shrewd business, as a direct payback from lightening the burden on biodiversity.

Most measures that avert waste have benefits for biodiversity, as the business footprint is lightened automatically at both the procurement and disposal stages.

4.3.6   So what can my business do?

Many of the activities of business are surprisingly close to biodiversity, resting on it to support key activities. With this come all of the risks outlined in the preceding chapter for businesses close to nature. Although it is common business practice to externalise both their costs and threats, it is just plain poor risk management to ignore the 'metabolism' of natural substances and products through the business. Nor is it sensible to have no concern for wildlife and landscape enjoyed by potential customers. Therefore, the prudent manager will progressively recognise these secondary dependencies on biodiversity, and begin to support appropriate biodiversity conservation efforts and/or seek sustainable use patterns for the long-term well-being of the business. We can once again discuss this in stages.

First, it is a matter of recognising that the business has fundamental dependencies on biodiversity and the services that it provides, despite common perceptions to the contrary.

Second, where specific elements of biodiversity are critical to use or enjoyment of products and services, such as healthy rivers to support the use of fishing equipment or outdoor clothing, the company should see what it can practically do to further conservation efforts of that landscape, habitat, species or other element of biodiversity.

Third, one has to map and, where possible, quantify the rates at which the business (or business activity) metabolises biodiversity, ecosystem services and other related resources.

Fourth, prudent risk analysis is required to determine where the demands of the business may place excessive pressure on the natural world, which may in turn threaten long-term supply chain and resource security.

Fifth, where possible, resources should be sourced from credibly certified sources. Despite the imperfections of the MSC and FSC marques, they represent a massive stride forwards from the kinds of unconstrained exploitation that have seen so many ecosystems degraded or destroyed. Who can afford the press feeding frenzy that might ensue if the media find that a business is less than scrupulous in the environmental and social impacts of its supply chain?

Sixth, the committed business should look for and follow up the many opportunities to cut the costs of energy and resource use as well as waste generation that litter current business practices. These can be planned and phased in ways that not only lighten the footprint on biodiversity but also relate back directly to an improved financial bottom line.

Seventh, the business should not be afraid of using its purchasing power to require its suppliers to compete not merely on price but also on an auditable process, assuring that materials are not produced using methods and sources that degrade biodiversity. Business is in the driving seat, not a mere passenger of what the market decides to offer on the basis of 'cheapest at any cost'. Indeed, business need not even be prescriptive about how its suppliers must act, but can instead create for them an incentive for innovations by specifying a demonstration of reduced impacts on biodiversity as part of the invitation to tender. These may include a requirement for potential suppliers to demonstrate clear, workable and active corporate policies towards biodiversity. Problem ownership is the key, even if the impacts on biodiversity are at one step or more removed. We can be sure that, in the event of the media or 'green' pressure groups picking up on the contribution of procurement and resource use by business to the destruction of an ecosystem, it will not be immune from blame and reputation damage.

The eighth step is then to look not merely at more sustainable sourcing of resources, but also at their wise stewardship through manufacturing processes, progressively to eliminate waste. Where residual waste is unavoidable, how can it be traded not merely to generate a return, or cut disposal costs, but also to eliminate waste flows? Every gramme of waste generated is not only profit lost but also unwarranted pressure on productive ecosystems and a burden on them to absorb the waste.

The ninth step is about product stewardship, seeking to design products that are amenable to safe use and, at end-of-life, re-use, recycling and benign disposal that does not overburden ecosystems. The waste hierarchy summed up by the well-known phrase 'reduce–re-use–recycle' is also a formula for business benefit through costs averted and markets anticipated.

As a tenth step, these procedures and policies should be enshrined in a management system that measures and controls relevant aspects of impacts on biodiversity. (I'll turn to further benefits or reporting on such a transparent process in the next chapter.)

Many companies have found advantages in tackling biodiversity and wider sustainable development challenges ahead of the eventual need to comply with regulations. Not only are they better prepared for when the regulatory regimes enter force, but reductions can be found in not only direct costs but also indirect costs such as taxation. (As perceived leaders, some businesses may also have some influence on the shape of such regulations.) These leading companies also accrue benefits from sound resource stewardship, development of technologies that can then be patented, and thereafter be controlled, licensed or sold on, and recognition by others of their credibility, for example through perception among social responsibility investors and other potential shareholders. A company's ranking on indices such as the Dow Jones Sustainability Index [15] or FTSE-4Good [16] has real significance for corporate value.

There are many more practical stories that could be told of how pioneering companies have generated real value in the real world from a responsible recognition and response to biodiversity issues. However, the above examples should suffice to illustrate both the breadth of opportunity and the hard bottom-line benefits that may be realised from taking an honest and far-sighted approach to biodiversity.

References

[1] Hardy & Greys, available at www.hardygreys.com.
[2] Orvis, available at www.orvis.co.uk.
[3] The Riverfly Partnership, available at www.riverflies.org.
[4] The Wild Trout Trust, available at www.wildtrout.org.
[5] The Anglers' Conservation Association, available at www.a-c-a.org.
[6] British Association for Shooting and Conservation (BASC), available at: www.basc. org.uk.
[7] The Salmon and Trout Association, available at www.salmon-trout.org.
[8] Yeo Valley Organic, available at www.yeovalleyorganic.co.uk.
[9] The Woodland Trust, available at www.woodland-trust.org.uk.
[10] Green & Blacks, available at www.greenandblacks.com.
[11] Intergovernmental Panel on Climate Change, available at www.ipcc.ch.
[12] Intergovernmental Panel on Climate Change, *Summary for Policymakers of the Fourth Scientific Assessment of the Physical Science of Climate Change*, February 2007.
[13] The Pew Center's Business Environmental Leadership Council (BELC), available at www.pewclimate.org/companies_leading_the_way_belc.
[14] Defra (UK Department for Environment, Food and Rural Affairs), *Stern Review on the Economics of Climate Change*, October 2006.
[15] Dow Jones Sustainability Index, available at: www.sustainability-indexes.com.
[16] FTSE4Good, available at www.ftse.com/Indices/FTSE4Good_Index_Series/index. jsp.

4.4   More distant from nature

From the discussion in the previous chapters, particularly about business as an 'ecological' entity that both consumes and excretes resources, it will probably come as little surprise that this chapter starts with the assertion that all business activities ultimately rest on biodiversity. Take away biodiversity with all its goods and services and, no matter at how many steps removed we use them, the sustenance of the business is cut off. Extinction will ensue. Contrary to common perception, all business activities are only a short step away from biodiversity.

So it is incumbent on all businesses to evaluate their relationship with the ecosystems that allow them to operate, from which benefits and threats flow, and which can therefore potentially continue to sustain them in the long term if a mutually supportive relationship is sought.

4.4.1   Dimensions of dependence

Most business activities probably fall into this category of a perceived greater distance from biodiversity. Indeed, the financial and IT markets can seem on the surface to have little relevance to biodiversity, but the consequences of decisions in each sector have massive implication for resource use and outcomes for biodiversity. Certainly, most of our inherited economic driving forces spur us on to greater profit for its own sake, with little or no regard for the natural (or human) capital that supports it today and can sustain the activity in the future. The short-termism of both the expectation of return to shareholders and the electoral cycle is hardly helpful to long-term planning for a sustainable relationship with biodiversity.

Consequently, it is often hard for those who understand the implications for biodiversity, or else feel an unquantified responsibility towards it, to get to grips with the practicalities of internalising legitimate biodiversity concerns into business practice. However, six dimensions of thinking facilitate the process of gaining acceptance for the business dependency on biodiversity, and for shaping the kinds of practical actions that are consistent with an increasingly responsible attitude. These six dimensions are: (1) risk management; (2) resource productivity; (3) business influence; (4) maximising biodiversity on landholdings; (5) stakeholder relationships and (6) corporate culture. We'll discuss each in turn.

4.4.1.1   Risk management

We have, in fact, already looked at many of the implications of prudent and comprehensive risk management in the preceding two chapters. All businesses have a dependency, whether directly or at remove, on the basic goods and services provided by biodiversity. And, as also discussed in the previous chapter, their customers may depend on biodiversity when using the products or services of the business. The prudent director or manager will have regard to all risks impinging on the business, which may boost or blight its long-term stability. Awareness of biodiversity and the dependence of business on it merely add to this catalogue of prudent risk management. As we are well aware, shareholders, particularly large corporate investors such as pension companies and the rating agencies that steer their decisions, are becoming increasingly insistent about the management of all business risks.

The insurance business, too, makes a direct connection between biodiversity and business. Indeed, insurance implications have been one of the major driving forces over recent years to cement the significance of climate change as a wider business issue, addressing the risks of both causal emissions as well as the downstream consequences of a more extreme climate. Insurance and reinsurance businesses are alert to changing future pressures on the relationship of business and biodiversity.

Planning consents are also becoming more tricky, protracted and expensive to secure for a range of reasons pertinent to sustainable development pressures. Biodiversity implications feature among these, in terms of protection of species and habitats, the functions they perform, and continued public access to green spaces and unspoiled environments.

The UK government is also becoming increasingly insistent, though as yet is well short of introducing statutory obligations, that corporate reporting should include implications for the environment including its biodiversity. Though hardly the ideal exemplar in terms of how it manages its own infrastructure and massive purchasing power, the UK government certainly audits the nation's progress towards sustainable development using a range of 'sustainability indicators' which include biodiversity outcomes such as populations of woodland and farmland birds. This growing background momentum towards open reporting is certainly focusing the minds of investors, ratings agencies and other influential actors within the business environment to have due regard to the wider 'footprint' of enterprises, including dependence on biodiversity. Some leading UK companies, such as Unilever, Wessex Water, The Co-operative Bank, Hydro Polymers and United Utilities, include implications for biodiversity and the wider environment in their voluntary 'sustainability reporting'.

In short, for business to be successful in future, and not merely to become a short-term profit-taker that is unlikely to stay the course, it has to become a responsible and respected player in the markets in which it seeks to operate. This is no mere matter of altruism towards biodiversity and society; it is, in fact, a matter of prudent risk management.

4.4.1.2   Resource productivity

The concept of resource productivity is one that already has considerable political and business momentum, addressing potential gains in eco-efficiency across enterprises which increase the production output per unit of natural resource input. In essence, it is about increasing the productivity of natural resources such that more human services and well-being are produced using less raw materials and energy. This is a society-wide challenge, but the self-evident benefits accrue directly to businesses servicing these societal needs.

Since most resources used by business ultimately have a biological source, the efficiency with which all resources are used across the business has a massive implication for the overall footprint of that business on the natural world. A culture of efficient and productive use of resources across business automatically contributes to a more responsible relationship with biodiversity.

The European Union actively promotes a sustainable approach to natural resources. The concept of resource efficiency, leading to the balancing of demand and natural capacity, is also an explicit part of UK government policy. For example, the UK government's revised sustainable development strategy, *Securing the Future*, published on 7 March 2005, contains a whole chapter (Chapter 3) devoted to 'Sustainable Consumption and Production' [1]. This is headlined under the term 'One Planet Economy', emphasising that current European resource consumption habits would, if extrapolated to the global population, result in a clearly unsustainable footprint of three planets. (This is discussed in the chapter 'People and nature'.) Since we only have one planet available to support our needs, urgent action is clearly essential.

Chapter 4 of the government's sustainable development strategy document, *Confronting the Greatest Threat: Climate Change and Energy*, also has significant implications for addressing the dependence of society and business on the natural world. Climate change, after all, is a consequence of our emissions exceeding the natural assimilative limits of the world's ecosystems. For more discussion of the business implications of this, see the discussion in the chapter 'One step removed'.

4.4.1.3   Business influence

In the previous chapter, I have already highlighted the numerous direct and indirect values to business of 'treading more lightly' on biodiversity. I will not labour the points again here, save only to reiterate the central importance of the supply chain and waste streams in determining the impact of business on biodiversity. Often, business may feel impotent in developing a more sympathetic relationship with biodiversity due to constraints within the supply chain. However, the reality is that, with a little strategic thought, business is far from being a passenger and can use various aspects of its economic influence to maximise biodiversity protection along the supply chain with its range of associated benefits. Not the least of these are security of supply and protection or enhancement of reputation.

As we saw with the origins of MSC and FSC certification, business is certainly no passive player that has to accept what the market offers. Indeed, as demonstrated by organic food and SCaMP, it can actually drive government policy to generate lasting change across an industry sector that is advantageous to the championing business. Therefore, business is a powerful player, either as individual companies or, more effectively, in trade associations or other consortia, when it comes to embedding concerns and innovations with respect to biodiversity.

Furthermore, as also demonstrated by SCaMP, by the support of pro-conservation angling organisations by distilleries, and by several other examples, business may benefit by promoting sympathetic use of biodiversity which, in turn, rewards it through resource security. A number of water companies exemplify this approach of influencing other businesses, sometimes by positive incentives. For example, in the late 1990s, Wessex Water (the water service provider throughout much of Wiltshire, Dorset and Somerset) advertised a financial incentive for farmers to convert to organic agriculture in catchments vulnerable to nutrient and pesticide contamination, in anticipation of lower seepage of these substances to watercourses and groundwater through organic farming practices. Although the scheme had little or no uptake, it does at least demonstrate the value of the influencing power of business to produce better biodiversity and ecosystem health as a resource for the business. Perhaps the plainest examples of this are provided for water resources, forest products and marine fisheries as addressed in the chapter 'Close to nature'.

4.4.1.4   Maximising biodiversity on landholdings

In some ways, this may seem an old-fashioned approach to biodiversity. Indeed, it is, if we are merely 'gardening' to create a semblance of environmental credentials that masks underlying corporate inertia. For example, neatly tended grounds housing a pesticide-producing company without good product stewardship processes is not only a mockery, but the neat manicuring itself is also unlikely to be of great value to biodiversity or corporate trust.

To be credible, testable and valuable, and supported in a way that ensures habitat management actually works, the maximisation of landholdings for biodiversity gain has to come from corporate values and commitment. These corporate values will recognise both the virtues and responsibility of business to biodiversity to mitigate its own inevitable impacts, augment a suite of other company-wide initiatives for protection of biodiversity, and be a manifestation of a wider culture of responsible management.

All corporate landholdings require a degree of management, so it is unlikely to be prohibitively expensive to manage landscapes more sympathetically for ecosystems and natural processes. Biodiversity-friendly land management can enhance both the attractiveness and value of land, create an amenity for local residents that fosters support for the business, and go some way to offset impacts on biodiversity in other activities of the business including the land 'take' of buildings, car parks and other infrastructure. In fact, some features of industrial infrastructure may have specific biodiversity benefits that can be maximised: for example, roosting, nesting or perching sites for prey birds in buildings and other structures. Measures such as best practice in averting bird strikes on power lines crossing watercourses also have an important role in harmonising infrastructure with biodiversity.

As emphasised above, management for biodiversity entails more than just planting greenery. For habitat to be functional, it must include native species of trees, herbs and other flora appropriate to the landscape if it is to support rich insect, bird, microbial and other biota dependent on them. Where badgers, water voles, butterflies, passerine birds or other biota already use the site, appropriate habitat needs to be protected or enhanced in sympathy with their needs. With careful planning, ideally involving local expertise such as that of the UK's regional Wildlife Trusts [2] or other special-interest groups, this can be entirely sympathetic with aspects of commercial landscaping such as screening, fencing (including wet fencing), landscaping, amenity and meeting spaces, etc. It should not be assumed that wildlife-oriented land management need be more expensive; in practice, it may entail little more than changes in mowing, hedge-cutting, pruning or replanting regimes, the elimination of pesticides, choice of plant species and similar measures which may even reduce costs throughout the year.

A further benefit of enabling wildlife to co-exist with corporate enterprise is that ecosystem functions will also be enhanced, and these can have wider benefits. For example, land of value to biodiversity will allow water to percolate, certainly to a far greater extent than concreted and other impermeable surfaces, and the more natural flows of water through the landscape can attenuate flash flooding. Where water is slowed by vegetation, suspended contaminants will be more likely to be precipitated out, and dissolved and microbial pollutants will be treated by natural physico-chemical processes. Pollution risk is therefore reduced through natural processes, averting potential costs. These hydrological, chemical, amenity and biodiversity benefits have been formalized in SuDS, or Sustainable Drainage Systems, in which the landscape and aspects of the built environment are designed in sympathy with natural processes of land–water interaction. Design in sympathy with biodiversity can therefore yield far more than altruistic benefits, including not only the above factors but also wet fencing, educationally valuable resources and community relationships.

The timing of construction or maintenance of buildings and infrastructure is also highly germane to biodiversity. By avoiding certain potentially disruptive maintenance or construction activities at important periods in different habitats – for example, the spring nesting period for birds, over-wintering of invertebrates in ponds or bat hibernation in buildings or hollow trees – biodiversity can co-exist far more successfully with commercial properties. Also, in some cases, the business may avert prosecution and impact on reputation from inadvertent disturbance of some species.

Some companies have found that explicit commitments to biodiversity conservation on their landholdings, or that of those they influence, can be advantageous for the many reasons already outlined above. For example, Wessex Water, albeit a company whose core activities are close to biodiversity, launched its 'Wessex Water Biodiversity Action Plan' in 1998 in response to the UK's Biodiversity Action Plan (BAP). (The BAP itself was the UK's response to obligations under the Convention on Biological Biodiversity signed at the Rio de Janeiro Earth Summit in 1992.) Wessex Water's BAP was the first corporate BAP in the UK, comprising the three key elements of funding for third parties to improve regional biodiversity, work on landholdings, and reduction of impacts from new infrastructure. This strategy demonstrates a broader corporate attitude to biodiversity which exceeds mere 'gardening', as described above. However, it should coincidentally help Wessex Water comply with emerging legislation (such as the EU Water Framework Directive and the EU Strategic Environmental Assessment Directive), restore river habitat and quality, grow corporate trust, and build partnerships with different organisations.

Since then, Thames Water has also implemented a 'Biodiversity Strategy'. Running from 1999, it addresses the four criteria of management of land- and water-holdings, activities related to water management, partnership working, and overall corporate responsibility. Southern Water has also now developed a corporate BAP, as have Severn Trent Water, United Utilities and a range of other businesses.

4.4.1.5   Stakeholder relationships

This all adds up to implications for the overarching concept of 'corporate reputation'. And, though a diffuse concept that is hard to define, we know from 'Black Monday' in October 1987, the Enron collapse of 2003 and other significant events how a sliding market or corporate reputation can, at best, harm corporate value or, at worst, send it into freefall.

Stakeholders, from employees through to shareholders, local residents, suppliers and customers, want to know that the enterprise is well managed and does not put them at undue risk or undermine their environment or other interests.

SACK AND BURNING OF THE CAPITAL OF ZEBU

Prudent risk management of impacts on biodiversity and, where appropriate, investment in its protection and restoration can assure all of those stakeholders. Corporate trust is a fickle concept, but the implications of losing it may be massive.

4.4.1.6   Corporate culture

The internal stakeholders of the business are a particularly important group of people. These people are what corporate rhetoric often announces to be the 'core resource of the business'. Together with biodiversity, this is probably true. Therefore, corporate culture is central to corporate success.

As we know from landscaping, plantings, waterfalls and fountains, both outdoors and within shopping precincts and corporate headquarters, proximity to nature in one guise or another gives people a feeling of well-being. Witness the infamous fish tank in the dentist's waiting room. The Harvard biologist E.O. Wilson [3] coined the term 'biophilia' in 1984 to describe an innate human sensitivity to and need for contact with other living things, because we have co-evolved and co-existed in the closest relationship with the natural world for so many millennia. This concept is supported by numerous peer-reviewed studies demonstrating that patients recover quicker if they are exposed to greenery, even pictures of greenery, rather than a purely artificial environment. We tend to lose well-being and act increasingly out of step with natural harmonious processes when divorced from living systems. And, in the business context, where the atmosphere is conducive, employees and visitors become more relaxed, less confrontational and increasingly productive. Corporate culture focused entirely on short-term profit-taking, overlooking its role in society and interactions with the natural world, is likely to be perceived as cold, mechanistic and disconnected with the personal values of staff and many customers. This is all important if people and business units are to mesh together effectively, contributing to a positive attitude and culture. It may also be important for staff retention, satisfaction and optimal performance. By caring for biodiversity, biodiversity may reward us not merely through the many tangible mechanisms already discussed but also via emotional and even spiritual dimensions, which can as effectively influence the financial bottom line.

Businesses may also elect to support a culture of employee volunteering for local initiatives to promote biodiversity, within clear criteria of course, where this contributes to the corporate aims and ethos of sympathy with biodiversity. United Utilities is one of many major UK businesses that take this enlightened view, while companies such as Mitsubishi Corporation (UK) plc and Aviva plc are corporate partners of the charity Earthwatch which leads expeditions to promote biodiversity projects across the world. Indeed, organisations such as Earthwatch and the British Trust for Conservation Volunteers (BTCV) have active corporate partnership initiatives to foster these relationships [4, 5]. As for any other investment in corporate culture, this can have real benefits in terms of staff morale, loyalty and retention as well as corporate perception by the community.

Through being credible, transparent and effective in managing biodiversity impacts, a business also becomes trusted. This may help it become a partner of choice of other businesses or organisations espousing responsible practices or supply chain standards, and may also sometimes attract less scrutiny from regulators. The organisation then can become an even more influential player in helping society make progress, supporting biodiversity by means that are compatible with the creation of wealth.

4.4.2   Practical actions and advantages

The first and most important stage in this evolution of corporate culture is to come out of denial, and to do so with neither guilt nor recrimination. Awareness of modern society's impacts on biodiversity began to dawn on us only about 40 years ago. So, when seen from the perspective of the amount of time we have inhabited the planet, it is not really surprising how relatively unsophisticated we are at getting to grips with both a robust language and practical tools to account for our use of and responsibilities towards the natural world.

Biodiversity has simply not been a feature of corporate thinking over most of our industrial past, nor of society's broader self-perception. Before this, of course, we grew up as an indivisible part of nature and used it to meet our needs, at first ad hoc by hunting and gathering and then with increasing foresight through farming systems. Only recently did we develop an industrialised capability that was founded on liquidation of natural assets to generate unprecedented wealth, longevity, public health and material expectation, educational provision and many benefits besides. Our awareness of the unintended consequences of our actions, and their implications for our long-term well-being and prosperity, is a very recent phenomenon.

So we need not feel the stultifying pressure of guilt about the fact that we are only just finding our way. Better to wake up now and grasp the issue openly and honestly. If any business out there is concerned about being pilloried by 'greenies' for being open and honest about shortfalls, I suggest you can relax. No one today has all the answers, but at least we are at last addressing the questions. We are all on this journey to sustainable development together; we have a long way yet to pathfind and travel, so the ire of the green movement is reserved mainly for those who deny the issues rather than those who make a serious but inevitably imperfect stab at grappling with them!

There is, in reality, no great distance between any business activity and the underpinning resource of biodiversity. We just need to realise this and explore together how to act accordingly to realise the advantages of a responsible relationship between the two.

We will get better at it; indeed, we must do. We live in an interesting age, both in terms of society's broader choices about finding lifestyles that will not limit its options in the future and in the way that we translate this into practical day-to-day business decisions. From the issues discussed above, it is clear that there are many 'hooks' to realise corporate benefit from addressing biodiversity through 'enlightened self-interest'. We cannot, ultimately, ignore the growing need to act, either by compulsion or by forethought. It is clear which will give us strategic advantage.

4.4.3   So what can my business do?

The journey starts when we abandon denial or ignorance that the link between biodiversity and business is important. Unlike in the previous two chapters, I have not attempted a more detailed stepwise programme, based on practical examples, for these residual business activities seemingly more distant from nature. We actually have relatively few strong examples of initiatives applied to bring biodiversity considerations transparently and directly into the heart of decisions within business activities at some distance from nature. However, the six dimensions introduced and subsequently elaborated throughout this chapter – (1) risk management, (2) resource productivity, (3) business influence, (4) maximising biodiversity on landholdings, (5) stakeholder relationships and (6) corporate culture – provide a sounding board for discussion and exploration within business, and may then serve as a springboard from which to launch enquiry and innovative thinking that is germane to simultaneous biodiversity and business success.

Effective management of impacts on the natural world relies on our thinking strategically and innovatively about how best to deliver social benefits through wealth-creating activities that simultaneously maintain the integrity of biodiversity on which all our interests depend. Thoughts on how to embed this into a business are articulated in Part V.

References

[1] UK Government. *Securing the Future: UK Government Sustainable Development Strategy*, The Stationary Office: London. (http://www.sustainable-development.gov. uk/publications/uk-strategy/index.htm).

[2] Wildlife Trusts (regional and mainly organised on a county basis). www.wildlifetrusts. org.

[3] Wilson, E.O., *Biophilia: The Human Bond with Other Species*, Harvard University Press: Boston, MA, 1984.

[4] British Trust for Conservation Volunteers (BTCV), available at www.btcv.org.

[5] Earthwatch Europe, available at www.earthwatch.org/europe.

4.5   Empowering people and nature

No market is stable indefinitely. Markets change daily, with new technologies, manufacturing or producing bases, trade agreements, economic and regulatory instruments, consumer tastes and attitudes, and a host of other pressures. We have already looked at some of the diverse ways that biodiversity considerations will affect our familiar economy, and the ways the foresighted and responsible business can benefit from that.

However, this is only one of the world's economies, and an economy from which so much of the global population is excluded. The developed world has been so busy competing with itself – company to company, country to country, trading bloc to trading bloc, first world to second – that, if we have thoughts of the rest of the world's billions of inhabitants at all, we have viewed them as little more than an inconvenience and a source of cheap resources. Or even as a dumping ground for outmoded products, a growth market for as-yet unregulated products such as cigarettes, or a place to still trade in chemicals banned in our home nations.

4.5.1   A changing world view

Yet, slowly, we are learning to care. Whether because we are realising that the consequences of the substantive environmental problems facing our common future are increasingly global in nature, or because we have evolved sufficiently to develop a conscience that embraces all of our own species. Maybe the eye of the world's media is now shaming us into facing the pathetic reality of human misery that could erstwhile be sanitised under the charity-box banner of 'poverty'. Maybe the all-pervasive power of the internet forces wider awareness on us. Or perhaps we simply need new markets.

For whatever reasons, the rich world has moved from a position of overlooking the disadvantaged, through a paradigm of aid and support that imposed our own values on them, towards a new dawn of capacity-building to empower others. This is particularly so in relation to Africa. The new model of development aid is to build capacity through targeted expertise and funding to help people co-create their own means to lift themselves out of poverty, engaging in economic and social progress that is built on the durable foundations of their own dignity and needs.

Prime among these international commitments were those undertaken under the UN-mediated Millennium Development Goals as well as subsequent commitments by the G8 nations. Although many commentators have castigated the G8 for reneging on such commitments, particularly the renewed and strengthened aid commitments made at the 2005 and 2007 summits, the fact remains that some national governments at least are making substantive investments to empower the poor and formerly marginalised.

One of the underpinning principles of sustainable development is to assist people in finding their own pathways towards economic and social fulfilment and progress in ways that respect the limitations of often-harsh environments. The challenge is never greater than in arid environments, where water can be a limiting factor to a progressive and equitable society. This means rediscovering traditional land use methods and crafts and innovating new ones with less impact on the supporting environment to sustain people's livelihood choices.

Often, there is a huge equity issue to be addressed simultaneously within the broader sustainability challenge. Innovative technologies and dialogue are required to work with vested interests in these regions to help them become more efficient in the use of water, land and other limiting elements of the ecosystem such that space is made for those entering the economy. There are challenges for delivery of infrastructure that is cheap and durable, with new appropriate technologies serviced by efficient materials, and a growing market of innovative products to help people empower themselves without swamping the carrying capacity of environments that can support this.

4.5.2   Products and services sympathetic with biodiversity

There are real biodiversity challenges here. New products and services are required that are technically and resource-efficient, appropriate to the needs of local people and which enable them to live more fulfilled lives without eroding the biodiversity that supports such aspirations. After all, many of the world's 'biodiversity hotspots' (regions of particularly rich and diverse biological diversity yet simultaneously subject to the greatest development pressure) occur in places where people historically have not trodden the destructive industrial path that we have blazed, with all of its disregard for the integrity of nature.

And, before we assume that this applies only to our more disadvantaged brothers and sisters, we should reflect that those of us in the already-developed world have a lot of catching up to do. On the back of our particular pathway towards economic, health and social progress, we have already eroded a great deal of the biodiversity and associated ecosystem services on which future advancement may depend. After all, in a host-imperial world, it is not acceptable to go and 'discover' and 'civilise' another country that just happens to still be rich in the natural resources we have depleted at home! From increasingly efficient irrigation to engineering, transport to communication, the need for a lifestyle with supporting products and services that has less impact on biodiversity is a pressing challenge if we are to make room not only for others in the world but also for our own future.

So tomorrow is certainly full of opportunity for products and services that are more benign to nature. In fact, this will be one of the key predictable defining features in everything from cultivation to aviation, commercial fishing equipment to catering, toilets and taps to transport, and construction to conveyance. This will apply to services, too, from design and financial products through to bio-friendly telecommunications and beauty therapies.

4.5.3   Restoration of nature

But tomorrow's global market does not stop there, for we have at some point to seek to rebuild what we have lost. Nature provides the primary capital for so much of our business activity, not to mention societal well-being and overall 'quality of life'. We have inherited a damaged biosphere, and the pace of erosion of its supportive capacities has yet to abate. We are still eating the very natural capital that we need to sustain our collective tomorrows. It is time not merely to halt this decline, but actively to seek to rebuild what we can. Restoration of nature and its many functions may seem altruistic, but it is far from it. It is, in fact, the wisest investment we could ever hope to make in opening up opportunity for future business and human potential.

The prize for us all, whether already rich or emerging from poverty, is not merely to protect such biodiversity as remains, but to reach beyond to restore the harm already done so that the natural world can support a richer future of more varied opportunity for all.

Encouragingly, we are starting to do it already. The shoots may be small and scattered, but they are green and healthy. In the chapter addressing businesses, 'Close to nature', I reviewed the SCaMP programme, which is restoring habitat in the uplands of north-west England for the benefit of not merely wildlife itself but also the water storage and purification functions that it performs. Restoration of catchment functions under SCaMP and a range of other initiatives around the world shows us that protection and rehabilitation of biodiversity and its many processes can increase the capacity for future security and business opportunity. There are many other green shoots elsewhere. For example, public funds are released in South Africa to support the Working for Water programme, which pays local people to eradicate the alien vegetation that would otherwise accelerate the evaporation of water from water-stressed land. Through both of these mechanisms, public investment is ploughed into enhancement of the ecology and water efficiency of the landscape to build its resilience and capacity to support future human development. Across America, after the proud dam-building tradition of much of the twentieth century, many dams, both larger and smaller, impounding a number of major American rivers are now being removed to restore natural river functioning and the myriad benefits stifled from them by historic river-run impoundment.

Likewise, we are seeing some of the core principles of natural ecosystem processes restored in urban land through engineering methods known collectively as SuDS (Sustainable Drainage Systems, or 'source control' technologies as they are known in the United States). SuDS are an innovative and relatively low-maintenance means of handling flood risk at source by emulating natural catchment processes of water detention, slowing of flows, pollutant treatment and groundwater infiltration, through a set of principles applicable in both urban and rural areas. More sympathetic to nature than traditional piped drainage systems, some SuDS systems also provide amenity areas for local residents and may even develop valued assemblages of plants and animals. Slowly, ecosystem-centred technologies, more benign to biodiversity, are beginning to take root in our much-battered world where nature has historically been squeezed out by a prior blinkered model of economic and social 'progress'.

The practice of surrendering corners, margins or headlands of former 'green desert' intensive arable fields has also moved from 'eco-freak' niche into more of a mainstream activity. These 'ragged edges' restore not just birds and bugs but also their associated natural crop pest control and pollination services. Europe-wide, we are also reversing the decline in forest cover wrought over former centuries, and with it offering green spaces and amenity, purification of air in built-up areas, places close to people for wildlife, improved landscape and tourism values, and even improved residential and commercial property prices boosted by the contingent value of a perceived 'greener' environment.

4.5.4   Investing in our future

This entire and growing catalogue of reinvestment in nature requires products and services from business in an emerging market limited only by our imagination. There is even a growing market in consultancy services to help policy-makers realise and then plan for the opportunities of biodiversity-friendly lifestyles and technologies for the benefit of local, national and global society.

We are not apart from biodiversity; we *are* biodiversity. We live on the back of it and, whether we are African poor or European *nouveau riche*, we will live better and more profitably with more of nature to support us. The potential, simultaneously for both profit and the collective well-being of humanity, is ours to seize.

4.6 From action to strategy

So, we have looked at some of the things we can do to bring biodiversity into activities within the business, be they close to nature, one step removed or else seemingly remote from biodiversity. We've also looked at markets emerging from increasing societal awareness and concern for protection and/or restoration of biodiversity.

In reality, nothing that humans do is remote or without implication for biodiversity, so we must necessarily seek a means to integrate biodiversity concerns across the whole business in a strategic way. Failure to do so would bring with it the risks already outlined, which can mean resource scarcity and cost, reputation, difficulty securing development planning approval, sales lost, declining staff morale and retention, loss of investor confidence and a wide range of other issues.

In Part V, I'll discuss how to set biodiversity challenges into a broader context that helps us integrate it, along with social factors, right across a business.

Part V Biodiversity and strategy

5.1 The business of biodiversity

It is easy for biodiversity and sustainability scientists, activists and pressure groups to underestimate the pressures that business leaders are experiencing. There are pressing expectations of returns to shareholders, novel and existing regulations that address a fragmented array of issues, and narrowly prescribed financial and other personal performance targets. The exigencies of the short term can be overwhelming. And, of course, all of this takes place within the ever-tightening grip of competition, never greater in an increasingly globalised world.

5.1.1 Price is king

Also, let's not kid ourselves that we are competing in a world where cutting corners, slashing costs, oversight of impacts on nature and people, and the prioritisation of immediate profit over long-term security, can all yield short-term wins. Furthermore, those economic and governance systems inherited from the Industrial Revolution still remain substantially unreconstructed (see Part I), and bear down on us constantly.

And then you confront the difficulty of trying to make proposals, judgements and decisions within an established business culture in which long-term implications for biodiversity and society may be alien concepts, generally viewed as expensive altruistic gestures of no central relevance to the organisation. It is not a matter of bad intent within organisations, but rather the legacy and habits of business culture born in a prior era of ignorance of the importance of biodiversity. After all, our Victorian forebears thought that nature had an unlimited potential to supply resources and absorb wastes. As a consequence, it is hard to open the door to biodiversity and sustainability, regardless of the strength of the business case. This is a challenge demanding leadership and influence.

While the preceding pages outline many dimensions of the business case for integration of biodiversity into corporate behaviour, the power of rational and moral argument alone is not automatically compelling to all players in an organisation. It is a matter of generating a common view of beneficial end-goals and the timing of actions appropriate to realising them; it is also a matter of culture change. This book should help you do this within the unique culture and dependencies of your business.

5.1.2 Strategy and planning

Some short-term returns are certainly possible from addressing biodiversity matters, particularly around the resource efficiency issues highlighted in Part IV. These 'quick wins' are important to set you on the longer road towards building wider support across business management. In fact, environmental advisors (such as Envirowise, the UK government-sponsored free environmental advisor to business [1]) and sustainability advocacy charities (such as Forum for the Future [2]) continually express surprise that organisations seem reluctant to break with habits and realise the often substantial and immediate gains for the financial bottom line that can arise from simple measures that save on energy, water and other resources, and reduce waste.

However, it is also true that a greater part of the challenge of bringing biodiversity concerns into the heart of business decisions is because the benefits are likely to be realised only over a longer time frame. Some measures, such as river catchment rehabilitation to yield more reliable flows of better-quality water, may be uncertain in outcome and take a significant amount of time before returns on investment can be anticipated. How then does one integrate long-term aspirations into the short-termism that drives cycles of business decision-making, and often also the mind-sets of industry's regulators who can constrain the freedom of businesses to offer long-term leadership?

This apparent conflict is resolved through the implementation of strategy. The making of money is just one of the things that business does. It is an important outcome but, as we've seen from the reputation issues addressed previously, it is not the be-all-and-end-all. Successful businesses prosper and endure because they organise themselves around (as distinct from hide behind) clear mission statements that define their broader role in society. For example, enterprises may aspire to provide utility services that make life better for people, supply digital communications infrastructure for connecting people or offer society educational and entertainment opportunities. This is all a whole lot more inspiring than merely seeking to take people's money by whatever means! The corporate relationship with biodiversity is part of that culture-building vision of a business here to stay and deliver responsibly for the benefit of society.

This view of identity and goal, then, contextualises the long-term purpose of the business, into which incremental short-term decisions may fit. There are those things on which business may act in the short term, those in the medium term for which it can prepare the ground, and then those long-term intentions towards which incremental steps may be identified that are profitable in the shorter term. It is dangerous to merely 'park' these long-term aspirations in the 'too difficult' file, as it is often these that not only define much of the mission of an organisation but may also be the most powerful stimuli for innovation. Some, obviously, entail gaining the support and collaboration of a range of partners, which may include potential competitors and trade associations, suppliers and supply chains, customers, shareholders, regulators, local residents and employees, marshalled around a clear vision of a mutually beneficial future. We know where we are headed, which can frame our incremental decisions, and this allows us to be proactive in the way we create and react to opportunities that arise along the way.

This strategic analysis is relevant to all decisions within business, including but not restricted to those pertaining to biodiversity. Nevertheless, the relationship

with biodiversity has to be a core element of concern for the business that wants to stay in business in the longer term.

These are again fine words, but they are likely to wither under the cut-and-thrust of decision-making in the boardroom or shop floor if we don't come to regard them as part of the wider culture change towards a more sustainable model of business. Profits tomorrow depend on the steps we take today to protect biodiversity and other key resources. It is about corporate mind-set and value systems as much as it is about logic alone.

And, as we know, culture change is a slow process, and has to be if it is to stick. Influencing key players across management and elsewhere in the business, then, is one of those short- to medium-term activities that build slowly but incrementally towards a long-term strategy of integrating biodiversity considerations across the business. And, as long experience tells us, the most durable changes we make in an organisation are those co-created around a central vision, rather than imposed top-down.

5.1.3   The tough get going

The reality remains that business is tough. However, by informing ourselves of the significance of biodiversity and taking a strategic view of how to implement biodiversity considerations across the diverse activities that we undertake – be they close to nature, one step removed, more remote from biodiversity or the new opportunities of a changing world – this challenge becomes tractable.

Furthermore, as we slowly infuse and enthuse the culture of the business, a responsible relationship with biodiversity will increasingly become the obvious and right direction in which to head. It may also be a powerful locus of innovation within the organisation. After all, the business of biodiversity is an opportunity for all of us to stick around in the longer term.

References

[1] Envirowise, available at: www.envirowise.gov.uk.
[2] Forum for the Future, available at: www.forumforthefuture.org.uk.

5.2   The business of sustainability

If you've stuck with me so far, you will have gathered that all of business, as indeed all human ventures, ultimately rests on biodiversity. We erode biodiversity at our peril. A sustainable relationship with biodiversity, then, becomes far more than a matter of altruism. We seek it because it represents a key element of prudent risk management as well as investment in resource security, corporate reputation, culture and moral obligation. It is part of innovating and steering a collective future for all of society, and part of the greater purpose for which a business exists.

5.2.1   Our place in society

Exactly the same type of argument can be mounted for the relationship of business with society. Business is, after all, merely the method used in a capitalist culture to convert natural and human resources into products and services, which in turn satisfy human needs and wants. Cultural habits change over time, and will – we can be assured – continue to respond to shifting knowledge and attitudes in the future.

For example, we used to deliver products in part on the back of slavery, but this is frowned upon today (though, worryingly, the practice is far from dead in developing nations and even, reportedly, in the 'black economy' of the UK). Child labour along supply chains has come under increasing scrutiny over recent years, and is now widely rejected in most developed-world markets. For those multinational companies that do not easily accept the moral argument to respond to these changes in awareness and attitude, shifting consumer pressure has generally been adequate to enforce a change in procurement strategy and supply chain audit. In the era of the internet, more intrusive media and growing legal requirements of transparency, hiding from scrutiny and accountability is an increasingly remote option. Social attitude shifts perpetually, and will continue to add to the pressures on business.

From Fair Trade to Responsible Care, Forest Stewardship Council to corporate social responsibility, implications for a wider range of stakeholders, from neighbours to consumers, feature increasingly prominently in corporate concerns, regulation and brand identity. We will leave the social dimension of business there for now, for it is not the main thrust of this book. However, let's close this part of the discussion by noting that social considerations are also important elements of corporate strategy, joining responsibility for biodiversity as another key element of corporate trust and representing a dimension to which shareholders are becoming increasingly alert.

5.2.2 Integrating economy, ecology and society

This all, of course, co-exists within the economic imperative of business. The imperative to show a return on investment today may be pressing, but we are becoming increasingly sophisticated in our thinking about the need for stable investment and growth for more secure profits tomorrow. In part, this comes about through recognition that making money is just one of the things that business does. It has also to deliver on a wider suite of parameters if it is to fulfil its role in society. And, flowing the other way, the 'licence to operate' granted by society to a business rests on more than just the business progressively emptying society's pockets of cash!

So, long-term economic success is intimately linked with delivery of social progress and stewardship of underpinning biodiversity resources, as the well-being and future prosperity of society ultimately rest on biodiversity and functional ecosystems.

This interlinked virtuous triad of economy, society and ecology is, of course, a cornerstone concept of sustainable development, the outcome of international thinking through the 1970s and 1980s.

5.2.3 Embedding biodiversity into corporate strategy

So, then, after tackling different aspects of the biodiversity jigsaw, we are left with the conclusion that the natural world and its living organisms and processes underpin every corner of every corporate body. How then does business respond to biodiversity in a coherent, strategic and beneficial way?

We can no longer regard biodiversity as external to our actions and decisions. We could proceed not to bother to question the implications for the natural and human worlds of our corporate decisions but to use some of our profits to pay for some conservation scheme. This is, granted, a gross example, and one that the World Business Council for Sustainable Development (WBCSD, a consortium of around 180 international companies from more than 30 countries and 20 major industrial sectors with a shared commitment to sustainable development) recognised as widespread some years ago. Indeed, so widespread was it that the WBCSD coined for it the delightful acronym FROG – 'First Raise Our Growth' – describing a self-defeating spiral wherein businesses pay for a little conservation gain on the back of profits gleaned from liquidating a lot more of the natural world [1]!

Yet, gross though this analogy is, the reality is that many businesses today still, mainly inadvertently, take an 'outside-in' view of the world. The pressures on businesses today have never been more intense, due to faster product development times, 'just in time' manufacture, increasingly stringent regulations, corporate and international competition and so forth. So the headroom available to think about implications for the natural world, as indeed any 'futures' issues, is often hard to find.

It is better then to hard-wire implications for biodiversity into the core business model and values so that it permeates all business considerations automatically and integrally, rather than peripherally or retrospectively.

5.2.4   Becoming sustainable

After all, we just want our business to survive and grow. And we have somehow to achieve this in a fast-changing world, where the pressures from all sides are growing by the hour.

The concept of sustainable development is neither vague nor altruistic, nor born of wishy-washy green aspirations. It is in fact a firm, challenging and integrated concept born from interdisciplinary scientific roots that reflect pressures on the natural and human worlds, the implications of eroding natural supportive capacity, and the simultaneous achievement of economic and social progress within environmental limits. In other words, sustainable development is a powerful integrating concept, and one that cuts to the underpinning principles of what drives many of the diverse and seemingly disconnected issues confronting business in the changing world.

The very word 'sustainable', let us remind ourselves, is essentially about the capacity to continue indefinitely. And for a business to continue it must simultaneously deliver economic progress and human utility while protecting, and ultimately restoring, the biodiversity and other natural resources on which it rests.

5.2.5   Getting to grips

The trouble is that, since the concept of sustainable development reached public consciousness with the 'Brundtland Report' in 1987, literally hundreds of subsidiary definitions have been published. Many of these definitions distort the purpose and purity of the original idea, seeking to justify and perpetuate the historic vested interests that created the environmental and social catastrophes that necessitated the establishment of the concept in the first place. So we have first to be sure that we find a model of sustainable development that at once embodies the core principles of sustainable development – particularly the achievement of simultaneous economic, environmental and social progress – in a framework that is amenable to pragmatic business planning. When we have done so, we can draw biodiversity into the heart of decision-making where it belongs.

There are a few such models. For example, 'triple bottom line' accounting, advanced by John Elkington (from the strategic management consultancy and think-tank SustainAbility Ltd [2]), builds on the three key aspects of sustainable development: economy, ecology and society. For each, it proposes the same kind of accounting as is ubiquitous for an organisation's finances. Some find this tool helpful in conceptualising sustainability challenges, identifying exposures and making progress with implementing measures within their enterprises.

There are many more models for applying sustainable development within the corporate or organisational context. For example, the UK's 2005 sustainable development strategy, *Securing the Future,* highlights different strands of the challenge that may be addressed in parallel [3]. However, in assessing the value and rigour of the growing number of solutions offered for applying biodiversity and wider social and environmental factors to sustainable business, the manager should be selective. For example, if the method does not relate to the finite limits of environmental goods and services, it is deficient as it ignores the full implication of sustainable development and will not therefore ultimately help us take advantage of new opportunities or avoid pitfalls. Furthermore, some purportedly 'sustainable development' methods fail to recognise the core principle of the simultaneous realisation of social, environmental and economic progress, so anything that reeks of a trade-off between these factors in the name of 'sustainable' decision-making should be instantly rejected. (For example, one often hears that sustainable development decisions have been taken into account in a new commercial development where economic and social, largely employment-based, considerations have been found to outweigh the environmental element; this is a trade-off and not necessarily a sustainable decision which would seek innovation for simultaneous benefit to all three strands.) With these few characteristics in mind, the manager can become discerning. For this reason, I will not highlight the shortfalls among the many corporate 'sustainable development' tools available today. But I will highlight two specific frameworks that are of particular value as they are at once scientifically based, reflect the breadth of issues embraced by the concept of sustainable development (critically including biodiversity), and are developed for use within business and other organisational practices.

In my experience, the most robust and generally applicable model supporting practical sustainable development is The Natural Step Framework. The TNS Framework, promoted by international sustainable development charity The Natural Step [4], is a science-based set of tools for practical sustainable development planning and action, based on the principles by which the natural cycles of this planet are maintained. It also includes societal dimensions, with particular reference to equity, and comprises a set of tools ideally adapted to the business environment and the achievement of sustainable business including short-term as well as long-term profit. The overarching scientific model is, however, rooted in cycling of resources through the biosphere, recognising the primacy of biodiversity and its functions in determining exactly what is and what is not sustainable.

Another helpful and robust tool is the Five Capitals model. This model, promoted by the UK solutions-based sustainable development charity Forum for the Future [5], identifies the five principal, non-substitutable 'capitals' on which sustainable human progress depends: natural, social, organisational, manufactured and financial. It then treats them all as we treat financial capital today within our capitalist system, creating a set of accounts for the 'capital' and 'interest' of each type. Collectively, these set the bounds of what could comprise a sustainable enterprise, product or process. When sustainable, the 'capital' is not being eroded but the enterprise is trading on the 'interest' generated by the system. Within the Five Capitals model, natural capital is rightly highlighted as non-substitutable with other forms of capital.

There are other models and tools out there addressing discrete aspects of the bigger challenge of sustainable development: life-cycle analysis, ISO 14001, EMAS, AA1000, and many more. Many of these may be used as valuable tools for discrete purposes within the overarching strategic approach to sustainable development enabled by the TNS Framework or the Five Capitals approaches. However, the key consideration here is that powerful and effective approaches to addressing sustainable development in an integrated way must bring biodiversity considerations and human well-being into the same conceptual framework as the other priorities with which companies are already familiar in their strategic and operational decision-making. Also, importantly, they must support corporate decision-making that is profitable, which benefits from self-interest, and which allows for short-term decisions to be developed as incremental steps towards long-term strategic goals.

Humanity has evolved as one element of the complex ecosystems of this planet, and remains wholly dependent on those systems. It therefore follows that:

- If humanity cannot co-exist with nature then humanity can have no future;
- If an organisation cannot manage its impacts on the natural world, protecting or restoring the habitats and species with which it interacts, then it is creating for itself liabilities for the future;
- Working towards a mutually sustaining relationship with the natural world is the bottom rung of the ladder of sustainable development, on which all other aspects of human well-being and equity are dependent and
- If this reflects the minimum criteria for basic survival, more can be done to increase the capacity of business and society to operate in future. Restoration of biodiversity is a strategic investment by society and business to protect or enhance its opportunities for the future.

5.2.6   Biodiversity and sustainability

The reality of biodiversity and business is that each is a single dimension of a complex and interconnected world. Humanity and all its activities play within a theatre forged by evolution over the past few billion years, in which all of the fundamental processes and resources that we depend on are part of the functions of nature. Business is just one of many activities that humans engage in to participate in the satisfaction of needs and desires. It is the way that people in capitalist societies use human innovation and efforts to convert biologically derived goods and services into products with utility and financial value.

Modern concepts of sustainable development tackle these elements within the broader systems of which they are part. It has been hard, throughout the earlier sections of this book, to refrain from diving straight into wider discussion about sustainable development and society at large. However, I have stuck to the task of highlighting the narrower axis between biodiversity and business, which still poses significant challenges with which we all need to grapple. This axis between business and biodiversity is of fundamental importance, yet in many modern conceptualisations of sustainable development, biodiversity seems not to be recognised as of prime significance. Redressing this balance is the major rationale for this book.

Keeping the thinking narrower, yet having regard for the broader world with which it interacts, has been rewarding. It has brought a great deal of state-of-the-art thinking about sustainability to bear on the narrower concept of biodiversity, which is about so much more than altruistic protection of species and habitats. And, in turn, a detailed consideration of the relationship of business with biodiversity has illuminated aspects of the broader challenge of sustainable development. After all, as I keep reiterating like an old stuck record, it is biodiversity that ultimately under- pins all human well-being, economic potential and 'quality of life'.

References

[1] World Business Council for Sustainable Development (WBCSD), available at: www. wbcsd.org.
[2] SustainAbility Ltd, available at: www.sustainability.com.
[3] UK Government, *Securing the Future: UK Government Sustainable Development Strategy.* The Stationary Office: London (http://www.sustainable-development.gov. uk/publications/uk-strategy/index.htm).
[4] The Natural Step, available at: www.naturalstep.org.
[5] Forum for the Future, available at: www.forumforthefuture.org.uk.

5.3   Restoration

From a big conceptual perspective, we live in a significantly impoverished world due to historic and ongoing degradation of biodiversity. We have therefore grown up with a diminished and fast-declining capacity for humans to meet their potential. There is just a lot less 'nature' around us, at home and across the globe, than there has been throughout the whole of human history. There are also a lot more of us wanting a piece of it.

5.3.1   Squeezed by history

This means that society is being increasingly constrained in terms of the things we need – food, fresh water and air, quiet places, educational resources, economic goods and so forth – by the legacy and continuing trend of ecological destruction. Every day, every instant, we erode even more deeply the capacity of nature to service our needs and demands in the future.

Halting the trend is an absolute minimum requirement for survival. Reversing it is then a duty, for both moral and practical reasons, if we are to restore the headroom for business and society to meet its needs in the future. We owe that as our bequest to future generations.

5.3.2   Sustainable lives

It is more than just the apocalyptic fears of an ageing 'green campaigner' insisting that we are very far from a sustainable society. However, let's imagine for a moment that we have witnessed a profound cultural revolution across the globe such that society has ceased the rapid, often irreversible degradation of biodiversity. To achieve this, we will have dramatically reduced resource demands, changed consumption habits and achieved a balance with nature's capacity to supply those needs. This would represent the attainment of the state of sustainability, towards which so much of the progressive rhetoric of society points, albeit currently with a pronounced lack of proportionate practical commitment. Let's just perform a thought experiment that imagines us in that state.

How good is life? Will it be fun, and what are our reasonable life expectations and day-to-day experiences?

In reality, our lives may not look hugely different to how they are today. We still have to work on our contribution to the functioning of society, with the financial and status rewards flowing back to us that enable us to afford those things that we determine are of value to our lives. We still need schools and colleges, and we still suffer the tyranny of the alarm clock. We still have to shop for our requisites, although most of the food and other products we buy have necessarily travelled a shorter distance across the planet to reach us. Indeed, some may be very local and, who knows, might be traded in novel ways.

We still need energy systems, though these too will inevitably be more decentralised and renewable than today, and we'll still require the skilled professionals to maintain them and ensure revenues go to the right places to keep the whole infrastructure working. Also, we'll need builders, water engineers, urban planners and, in all probability (Heaven help us all!), lawyers, bankers and pension companies.

Although health and healthy lifestyles will doubtless be integral to the planning of communities, we'll all need doctors, hospitals and healthcare professionals too, besides strong governance systems, if we are to ensure that our material aspirations remain within the carrying capacity of biodiversity and that this is equitably shared across society. Sad to say, we will probably still live in a society in which that political system, from parish through district and country to national, is abused by certain individuals and cliques who put self-interest first. However, the checks and balances will – let's be optimistic – work a little better than they do today and the overall political machine will be rather more coloured by the aspiration of a sustainable future than it is in its current incarnation.

Given that we are unlikely to have invented a more pervasive model than the capitalist system, let's assume we'll be living with that capitalist system, albeit modified to reflect more of the facets of a sustainable society. The current model is not entirely inconsistent with that new aim, if we iron out of it the perversities and build in that which is currently externalised (mainly the all-important facets of biodiversity and other environmental resources as well as impacts on people) [1]. So we will probably still be living with disparities of wealth and power, although the more gross skews we see today will have been tweaked with a specific intent to bring the currently poor and marginalised into the mainstream.

It is hardly a perfect world, but at least one with a viable future. We'll still have our ups and downs, drudgery and escapism, holidays, annoying neighbours, ice cream in hot weather, taxes, magazines, DIY stores and office stationery suppliers. But we'll also at least have a long-term future.

And we'll live closer to weeds, bugs and other creatures not because we have reverted to living in caves, but because we have planned to a far greater extent for co-existence with biodiversity and natural processes. The fabric of our commercial and residential areas (which are likely to be far better integrated than today) will have been planned to permit more natural fluxes of water and geochemicals, resident and migratory species, and contact with green spaces and their biota. We'll still have nature reserves for the most endangered wildlife and critical ecosystem functions, and also to ensure some areas of suitable scale for the life requirements of different species and ecosystems. Some people will still get hay fever. However, nature and people will live closer because we see the health, hydrological, psychological, economic and myriad other cases for this to be so.

But how good will life be? Well, that will be limited, entirely and unavoidably, by the carrying capacity of biodiversity.

5.3.3   In balance

Whatever the character of future sustainable society, a defining feature will be that it will be bounded by the finite carrying capacity of the ecosystems on which it depends. This type of thinking will be automatic for most people in future, except that we will have moved beyond today's clumsy terminology. Balance with nature will be as natural to us, whatever the words we use to articulate it, as other, more widely agreed aspects of 'justice', 'responsibility' and 'fairness' are today. As a society, we will have internalised it as a completely embedded feature of our common consciousness, legislative system and the economic pushes and pulls that flow from it.

   The trouble is: we are the product of our history. And, as is abundantly clear, the biodiversity that we have inherited is substantially damaged. If we continue to live within a badly degraded system, we will be living at a lesser potential than would be the case were our biodiversity intact and our ecosystems functioning fully.

5.3.4   Restoration

The restoration of habitat may today be firmly in the bracket of altruism, perhaps even of nostalgia, as if returning habitats to a perceived historic better state were merely 'nice to have'. In fact, returning them thus is probably a practical improbability. First, any historic reference point is, by very definition, arbitrary. Its selection will be coloured by myriad value judgements, shortfalls in knowledge and personal preferences. Second, if it were to be sustained, we would have to recreate the diverse driving forces that shaped a former status of biodiversity that way in the first place. Third, nature is nowhere deterministic, as the direction of ecosystem succession is essentially chaotic and, anyhow, a 'seed bank' of all relevant biological organisms may no longer be freely available in the environment. Fourth, the unpredictable implications of climate change introduce yet another 'wild card' about the direction that biodiversity will take when we redress other pressures on the environment.

   Yet, while a carbon copy of yesterday's habitat is unlikely and not necessarily wholly desirable, restoration of the unimpacted conditions of the habitat and its suitability for a similar type of ecosystem is both feasible and desirable. It is certainly invaluable if we want to restore the quality and quantity of ecosystem functioning to a less damaged, more resilient and supportive condition. It is, after all, this reconstructed ecosystem functionality that provides the bedrock of society's needs and aspirations through the diverse ecosystem services it 'produces'.

   Our conception of 'quality of life' would then need to be coloured not so much as today by more material goods, but by more intact and functionally efficient ecosystems that can provide for society a greater headroom of resources to meet its needs. Restored catchments will yield greater and more reliable flows of clean water and support more diverse fisheries, wildfowl and other biological resources, have a more natural flood regime and also naturally fertilise flat lands from which more goods may be harvested with less destructive tillage and inputs of agrochemicals and petrochemicals. This will provide more habitat for amenity, education and quiet reflection, adding to the value of urban and industrial developments. It will build richer soils, less erosive and of higher humic content. An investment in natural capital, to improve the capacity of biodiversity and its many functions to support greater economic growth and other facets of human progress, is one of the most strategic decisions that society can make for realising its potential into the future.

This is not just desirable, but essential if we are to widen the scope of opportunity for ourselves and future generations. More and better timber, reliable clean water and air, richer commercial and recreational fisheries, increased natural beauty, more productive soils, biomass for energy production and construction needs, greater resilience, richer genetic resources, deeper heritage and so on and on.

All of these ecosystem services provide the means for greater life satisfaction and realisation of human potential: a better legacy, for sure, than handing down damaged goods that limit human potential and security, and which just as certainly inhibit economic opportunity.

5.3.5   Growing business

If we can restore aspects of the Earth's ecosystems, we can deliver sustainable improvement in the quality of people's lives and leave behind improved prospects for our collective future. With a greater wealth of biodiversity to service society's needs and tastes, the economic opportunities will also grow. Business is, as I keep reiterating, merely the way that society converts natural resources into useful products within a capitalist system. And where there are human needs and an improved resource base to service it, business can thrive. Indeed, as we have seen with the restoration of headwaters of river catchments and the protection of forests and marine fisheries, business may already be in the driving seat of enhancing biodiversity to improve business security in the future.

We need to act now to understand the place and importance of biodiversity in our world, seek to reduce our demands on its limited resource and realise some of the benefits of this 'enlightened self-interest'. Once we have done that, we can invest in restoring biodiversity's capacity to broaden options for human well-being and business into the long-term future.

Reference

[1] Porritt, J., *Capitalism as if the World Matters*, Earthscan Publications: London, 2006.

6.1 Epilogue

It is my sincere hope that this book will very rapidly become out of date. This is not because I believe the issues it addresses are in any way time-limited. Biodiversity has always underpinned the inescapably ecological nature of human well-being, wealth-creation activities and 'quality of life'. Furthermore, as we become increasingly sophisticated in our approach to the wider concept of sustainable development, practical measures to address biodiversity considerations will become more rather than less important. Our perception of the fundamental importance of the relationship of business and society with the natural world must necessarily only deepen. It is just that we need to become more skilled at thinking about and acting on it.

Part of this is scientific. Today, we have little practical understanding of the way that nature produces many of its myriad beneficial ecosystem services including, for example water and air purification, habitat maintenance, climate regulation or the building of soil fertility. We need to understand far better how nature does these things so that we can more appropriately appreciate and value them in public and corporate decision-making, and be better placed to understand what constitutes truly sustainable resource use.

We have a significantly greater understanding of the capacity of nature to produce its various 'provisioning services' (often previously referred to as ecosystem 'goods') – timber, fin fish and shellfish, fresh water and so forth – and how to manipulate ecosystems to selectively boost productivity of economically valued resources through, for example farming, aquaculture and other methods. Yet today we translate this only poorly into decision-making, and we do not adequately account for the implications of boosting a particular service for our own ends (such as damming a river to boost water yield or ploughing virgin habitat for crop production) on overall ecosystem integrity and the wide range of other ecosystem services on which other people may depend. So we need better understanding and collaborative decision-making to facilitate the practical application of resource conservation in the boardroom and in government. Other ecosystem services provided by the biodiversity of natural ecosystem services, including for example floodwater detention by habitat within catchments or crop pollination through wild insects, are more difficult to implement in the boardroom, so we need to find mechanisms to address them through collaboration and adaptive forms of regional, national and international governance.

Today, we are, in short, unsophisticated in terms of basic knowledge and management of the interaction of biodiversity, business activities and other societal interests. In future, we need to be able to implement simple audits of the footprint of business decisions on biodiversity and social cohesion, and transparently reflect this in appropriate market prices, taxation, mitigation, prohibition or incentives to act in different ways. Biodiversity may be a common good for all of society, but we currently lack the ability to audit the way we draw down on it in our decisions and to build this equitably into a sustainable market mechanism.

We need this, not merely as an interesting intellectual activity, to be applied as a bespoke, altruistic activity by niche businesses but also as robust, commonly applicable tools, relevant and comparable across companies, business sectors and other types of organisations, so that we can address societal impacts on biodiversity on a level playing field. A sustainable relationship with the ecosystems that support us is, after all, mandatory if we are to cease to degrade the natural world or, better still, restore its functional capacity and broaden the options available to business and society in future.

But it is not just the technical end of things where we need to be innovative. We also need an appropriate language, the framing of that language in terms of opportunity and not just threat, the democratic will to act and innovations in collaborative decision-making and co-creation of solutions.

It is, for example quite within our grasp today to construct a cogent argument about how biodiversity underpins societal aspirations, and exactly why its protection and restoration is essential for a sustainable future. It is, however, far more difficult to persuade the lay person – say through a daily newspaper or a chat with friends in the pub – of exactly why the brown hairstreak butterfly in a local nature reserve, or more likely on an area of land threatened by development, really matters. Both the 'expert' and 'common parlance' examples concern the same thing, yet the pair highlight the complexities of translating concepts between scales, and the paucity of language today to achieve that. We are not there yet, clearly, but need to get there very soon.

We also need that appropriate language and 'stories of meaning' if we are to become effective at influencing politicians and others entrusted with the evolving public policy agenda within which businesses operate. We depend on this for sustainable governance, including the often-discussed yet currently remote 'level playing field' for wise decision-making.

Which brings us back to the business boardroom. Various arguments in favour of an 'enlightened self-interest' approach to biodiversity have been raised throughout this book, including sound stewardship, risk management, resource productivity, corporate reputation, staff morale, sustainable profit and so forth. However, sophisticated decision-making with respect to biodiversity, as for other aspects of sustainable development, needs to become not only second nature, but a consideration intrinsic to corporate decisions, culture and behaviour.

How do we know if we are making real progress with sustainable development? We have in biodiversity an unambiguous barometer of ecosystem health, offering us an objective measure of the benefits that ecosystems provide for us and all who share the planet. With it, we also have a 'barometer' of the health of our supportive ecosystems, and a way of knowing whether human activities and management practices are taking us in a sustainable direction or not. The choice to put them into effect, and the tools to do so, are in our hands.

Tomorrow's 'business as usual' will be – indeed needs to become – quite different to today's conception of 'usual' with all of its externalisations of biodiversity and social interactions.

In short, biodiversity itself needs to become natural to business thinking. One day, let's be optimistic, it will be entirely natural to value and protect the natural capital that makes business possible, beneficial to society and profitable.

The most important steps towards that goal are the ones that we take now.

Part VII   Appendix

Some definitions of biodiversity

The following definitions of biodiversity were culled from a trawl of the internet in late 2007. Each differs in detail, yet shares common elements of use to the understanding of biological diversity and its implications.

The variety of life on our planet, measurable as the variety within species, between species, and the variety of ecosystems.
NatureWatch: www.naturewatch.ca/english/plantwatch/dandelion/glossary.html)

Number and variety of living organisms; includes genetic diversity, species diversity, and ecological diversity
www.biology.usgs.gov/s+t/noframe/z999.htm

The number and variety of organisms within one region. This includes also the variability within and between species and within and between ecosystems.
www.o2.com/o2_glossary.asp

A large number and wide range of species of animals, plants, fungi, and microorganisms. Ecologically, wide biodiversity is conducive to the development of all species.
Natural Resources Defense Council: www.nrdc.org/reference/glossary/b.asp

The relative abundance and variety of plant and animal species and ecosystems within particular habitats.
www.harvestenergy.com/GlossaryPower.html

The number and variety of different organisms in the ecological complexes in which they naturally occur. Organisms are organized at many levels, ranging from complete ecosystems to the biochemical structures that are the molecular basis of heredity. Thus, the term encompasses different ecosystems, species, and genes that must be present for a healthy environment. A large number of species must characterize the food chain, representing multiple predator-prey relationships.
US National Safety Council: www.nsc.org/ehc/glossary.htm

The number and types of organisms in a region or environment.
The Pew Charitable Trusts: pewagbiotech.org/resources/glossary/

The variety and abundance of species, their genetic composition, and the natural communities, ecosystems, and landscapes in which they occur.
Washington Secretary of State Digital Archives: www.digitalarchives.wa.gov/ governorlocke/gsro/glossary.htm

The variability among living organisms from all sources including terrestrial, marine and other aquatic ecosystems and the ecological complexes of which they are part; this includes diversity within species, between species and of ecosystems.
Environment Canada, Species at Risk: www.speciesatrisk.gc.ca/glossary_e.cfm

The diversity, or variety, of plants, animals and other living things in a particular area or region. It encompasses habitat diversity, species diversity and genetic diversity.
East Herts District Council: http://www.eastherts.gov.uk/index.jsp?articleid=2608

The variety of life forms: the different plants, animals and micro-organisms, the genes they contain and the ecosystems they form. It is usually considered at three levels: genetic diversity, species diversity and ecosystem diversity.
www.tasmaniatogether.tas.gov.au/tastog_original/tt_glossary.html

The variety of life and its processes; it includes the variety of living organisms, the genetic differences among them, the communities and ecosystems in which they occur, and the ecological and evolutionary processes that keep them functioning, yet ever changing and adapting.
Saginaw Bay Greenways Collaborative: www.msu.edu/~jaroszjo/greenway/ glossary/glossary.htm

The genetic, species, and ecological richness of the organisms in a given area.
South Carolina Oyster Restoration and Enhancement Program (SCORE): www3.csc.noaa.gov/scoysters/html/glossary.htm

The variety of different species, the genetic variability of each species, and the variety of different ecosystems that they form.
Environment Canada: www.ec.gc.ca/water/en/info/gloss/e_gloss.htm

The variability among living organisms from all sources, including, among other things, terrestrial, marine and other aquatic ecosystems and the ecological complexes of which they are part.
FISHONLINE: www.fishonline.org/information/glossary/

Biological diversity; can be measured in terms of genetic, species, or ecosystem diversity.
Estrella Mountain Community College, Online Biology Book: www.emc. maricopa.edu/faculty/farabee/BIOBK/BioBookglossB.html

A shortening of the term 'biological diversity.' The diversity of life on Earth. The variability among living organisms and their interactions, both within species and between species, between ecosystems and across landscapes.
The Pacific Forest Trust: www.pacificforest.org/about/glossary.html

The variety and variability of life in an area or the diversity of genes, species, and ecosystems, and of plant and animal life within species (genetic diversity), among species (species diversity) and among ecosystems (ecosystem diversity). The latter includes the diversity of structure and function within ecosystems. The variety of life in all forms, levels and combinations. (IUCN) There are three types of biodiversity: ecosystem diversity, species diversity, and genetic diversity.
Take Back Wisconsin: www.takebackwisconsin.com/Documents/Glossary.htm

The variety, distribution and abundance of living organisms in an ecosystem. Maintaining biodiversity is believed to promote stability, sustainability and resilience of ecosystems.
University of Florida, School of Forest Resources and Conservation: www.sfrc.ufl.edu/Extension/ssfor11.htm

The variety of forms—the different plants, animals and micro-organisms, the genes they contain, and the ecosystems of which they form a part.
www.malleecma.vic.gov.au/glossarymcma.asp

The existence of a wide range of different types of organisms in a given place at a given time.
Queensland University of Technology, Student Web: www.students.ed.qut.edu.au/n2364379/MDB377/GlossaryPage.html

Biodiversity is the abundance of different plant and animal species found in an environment.
Enchanted Learning: www.enchantedlearning.com/subjects/butterfly/glossary/indexb.shtml

The full range of natural variety and variability within and among living organisms, and the ecological and environmental complexes in which they occur. It encompasses multiple levels of organization, including genes, species, communities and ecosystems.
The Nature Conservancy: www.nature.org/aboutus/howwework/cbd/science/art14307.html

The variety of species and ecosystems, the variability of genes within the species and the ecological complexes of which they are a part.
www.sustainableag.net/glossary_a-d.htm

The tendency in ecosystems, when undisturbed, to have a great variety of species forming a complex web of interactions.
Commonwealth of Massachusetts:
commpres.env.state.ma.us/content/glossary.asp

The diversity of plant and animal life in a particular habitat (or in the world as a whole); 'a high level of biodiversity is desirable'.
WordNet: http://wordnet.princeton.edu/perl/webwn?s=biodiversity&sub=Searc h+WordNet&o2=&o0=1&o7=&o5=&o1=1&o6=&o4=&o3=&h=

Biodiversity or biological diversity is the diversity of and in living nature. There are a number of definitions and measures of biodiversity.
Wikipedia: en.wikipedia.org/wiki/Biodiversity

Index

WITPRESS   *...for scientists by scientists*

Management of Natural Resources, Sustainable Development and Ecological Hazards II

Edited by: **C.A. BREBBIA**, *Wessex Institute of Technology, UK, and* **E. TIEZZI**, *University of Siena, Italy.*

The first Conference was very well attended by a substantial group of scientists from all over the world and helped to underline the concern of the international community regarding the state of the planet. The basic premise of the meeting was the need to determine urgent solutions before we reach a point of irreversibility.

Our current civilisation has fallen into a self-destructive process by which natural resources are consumed at an increasing rate. This process has now spread across the planet in search of further sources of energy and materials. The aggressiveness of this quest is such that the future of our planet is in the balance. The process is compounded by the pernicious effects of the resulting pollution.

On topics presented at the Second International Conference on Management of Natural Resources, Sustainable Devlopment and Ecological Hazards include the following headings: The Re-encounter; Political and Social Issues; Planning and Development; Safety; New Technologies; Energy; Training and Information; Learning from Nature; Ecology; Health Risks; Water Resources; Air; Soil.

WIT Transactions on Ecology and the Environment, Vol 127
ISBN: 978-1-84564-204-4   2009   apx 600pp
apx £198.00/US$396.00/€257.00
eISBN: 978-1-84564-381-2

WIT*Press*
Ashurst Lodge, Ashurst,
Southampton,
SO40 7AA, UK
Tel: 44 (0) 238 029 3223
Fax: 44 (0) 238 029 2853
E-Mail: witpress@witpress.com

 **WIT**PRESS *...for scientists by scientists*

Sustainable Development and Planning IV

Edited by: **C.A. BREBBIA**, *Wessex Institute of Technology, UK,* **M. NEOPHYTOU,** *University of Cyprus, Cyprus,* **E. BERIATOS,** *University of Thessaly, Greece,* **I. IOANNOU,** *University of Cyprus, Cyprus and* **A. G. KUNGOLOS,** *University of Thessaly, Greece*

The Conference addresses the subjects of regional development in an integrated way in accordance with the principles of sustainability. It has become apparent that planners, environmentalists, architects, engineers, policy makers and economists have to work together in order to ensure that planning and development can meet our present needs without compromising the ability of future generations.

Effective strategies for management should consider planning and regional development, two closely related disciplines, and emphasise the demand to handle these matters in an integrated way. This conference provides a common forum for all scientists specialising in the range of subjects included within sustainable development and planning.

Papers from the Fourth International Conference on Sustainable Development and Planning including tpapers from the following topics: Regional Planning; City Planning; Rural Development; Environmental Impact Assessment; Environmental Management; Environmental Legislation and Policy; Integrated Territorial and Environmental Risk Analysis; Ecosystems Analysis, Protection and Remediation; Social and Cultural Issues; Environmental Economics; Geo-Informatics; Urban Landscapes; Transportation; Waste Management; Resources Management; Forecasting; Politics and Sustainability; Sustainable Development in Developing Countries; Indicators of Sustainability; Response to World Events; Marine Environment.

WIT Transactions on Ecology and the Environment, Vol 120
ISBN: 978-1-84564-181-8   2009   apx 600pp
apx £195.00/US$390.00/€249.00
eISBN: 978-1-84564-358-4

 **WIT**PRESS   *...for scientists by scientists*

The Sustainable City V
Urban Regeneration and Sustainability

Edited by: **A. GOSPODINI**, *University of Thessaly,*
Greece, **C.A. BREBBIA**, *Wessex Institute of Technology,*
UK, and **E. TIEZZI**, *University of Siena, Italy*

A sustainable city is widely recognised as one that
meets the needs of the present without compromising
the ability of future generations to meet their own
needs. In The Sustainable City V many interrelated
aspects of the urban environment from transport and
mobility to social exclusions and crime prevention are
addressed. The papers included were originally presented at the Fifth
International Conference on Urban Regeneration and Sustainability and will
be of interest to city planners, architects, environmental engineers and all
academics, professionals and practitioners working in the wide range of
disciplines associated with creating a viable urban environment.

The papers are published under the following topics: Strategy and
Development; Planning, Development and Management; Environmental
Management; Planning Issues; Socio-Economic Issues; The Community and
the City; Cultural Heritage; Architectural Issues; Traffic and Transportation;
Land Use and Management; Public Safety; Sustainable Transportation and
Transport Integration; Energy Resources Systems; Healthy Cities; Urban-Rural
Relationships; Spatial Modelling; Mega Cities; Indicators: Ecological,
Economic, Social; Revitalisation Strategies.

WIT Transactions on Ecology and the Environment, Vol 117
ISBN: 978-1-84564-128-3   2008   768pp
£253.00/US$506.00/€329.00
eISBN: 978-1-84564-350-8

WIT*Press*
Ashurst Lodge, Ashurst,
Southampton,
SO40 7AA, UK.
Tel: 44 (0) 238 029 3223
Fax: 44 (0) 238 029 2853
E-Mail: witpress@witpress.com

 **WIT**PRESS   *...for scientists by scientists*

Sustainable Tourism III

Edited by: **F.D. PINEDA**, *Complutense University, Spain*
and **C.A. BREBBIA**, *Wessex Institute of Technology, UK*

Tourism, internationally, is the largest economic sector both in terms of earnings and number of people employed. Understandably, the economic advantages have led to the active promotion of tourism by governments and other institutions, often independent of the consequences on the environment. The challenge is to balance the need for a low impact on the environment and local culture, while helping to generate income, employment and the conservation of local ecosystems. Sustainable tourism has to be both ecologically and culturally sensitive.

This book contains papers presented at the Third International Conference on Sustainable Development, held in Malta. The Meeting focused on empirical work and case studies from around the world, and the book offers new insight and best practice guidance for supporting sustainable tourism. Adopting a multi-disciplinary approach, this book examines the practice of sustainable tourism from global travel trends through to destination and site management.

Of interest to scientists, practitioners and policy makers, the topics covered in this volume include: Climate Change and Tourism; Community Involvement; Risk and Safety; Rural Tourism; Tourism and Protected Areas; Tourism as a Factor of Development; Tourism Impact; Tourism Strategies.

WIT Transactions on Ecology and the Environment, Vol 115
ISBN: 978-1-84564-124-5   2008   368pp
£121.00/US$242.00/€157.50
eISBN: 978-1-84564-346-1

WIT eLibrary

Home of the Transactions of the Wessex Institute, the WIT electronic-library provides the international scientific community with immediate and permanent access to individual papers presented at WIT conferences. Visitors to the WIT eLibrary can freely browse and search abstracts of all papers in the collection before progressing to download their full text.

Visit the WIT eLibrary at
http://library.witpress.com